edexcel

Edexcel International Physics
Edexcel Certificate in Physics

Revision Guide

Steve Woolley

PEARSON

Published by Pearson Education Limited, a company incorporated in England and Wales, having its registered office at Edinburgh Gate, Harlow, Essex, CM20 2JE. Registered company number: 872828.

www.pearsonglobalschools.com

Edexcel is a registered trademark of Edexcel Limited

Text © Pearson Education Limited 2011

First published 2011

20 19 18 17 16 15 14 13 12

IMP 12 11 10 9 8 7

ISBN 978 0 435046 7 36

Copyright notice

Edited by Linnet Bruce

Proofread by Marion Edsall and Elizabeth Barker

Original design by Richard Ponsford

Typeset by Naranco Design & Editorial / Kim Hubbeling

Original illustrations © Pearson Education Ltd 2011

Illustrated by Andriy Yankovsky

Cover design and title page by Creative Monkey

Cover photo © Getty Images: Peter Dazeley

Printed in Italy by Rotolito Lombarda

Acknowledgements

The author and publisher would like to thank the following individuals and organisations for permission to reproduce photographs:

(Key: b-bottom; c-centre; l-left; r-right; t-top)

Alamy Images: 27c, 27b, 52tl, 53tl, 60; Author's own work: 18 (damaged plug), 18/a (plug), 18/b (Cable), 18/d (Kettle), 53br, 73 (Transformer); DK Images: 27t; Fotolia.com: 13cr, 15cl, 28t, 28c, 28b, 38/12.2, 52bl, 53/16.5, 60/10b, 73 (compass), 73 (magnets), 84br; iStockphoto: 9bc, 9br, 60/10a; Pearson Education Ltd: 19tr, 73/d; Science Photo Library Ltd: 24bc, 24br; Shutterstock.com: 9bl, 17tl/a, 17tc/b, 17c/d, 17cl/c, 18 (Safety plug), 19tc, 24c, 24cr, 38/12.1, 38/12.3, 38/12.4, 38/12.5, 38/12.6, 38/12.7, 52tc, 53/16.6, 55bl, 60/11, headers, title-page

Cover images: Getty Images: Peter Dazeley

All other images © Pearson Education 2011

Every effort has been made to contact copyright holders of material reproduced in this book. Any omissions will be rectified in subsequent printings if notice is given to the publishers.

Websites

The websites used in this book were correct and up to date at the time of publication. It is essential for tutors to preview each website before using it in class so as to ensure that the URL is still accurate, relevant and appropriate. We suggest that tutors bookmark useful websites and consider enabling students to access them through the school/college intranet.

Disclaimer

This material has been published on behalf of Edexcel and offers high-quality support for the delivery of Edexcel qualifications.

This does not mean that the material is essential to achieve any Edexcel qualification, nor does it mean that it is the only suitable material available to support any Edexcel qualification. Edexcel material will not be used verbatim in setting any Edexcel examination or assessment. Any resource lists produced by Edexcel shall include this and other appropriate resources.

Copies of official specifications for all Edexcel qualifications may be found on the Edexcel website: www.edexcel.com

Contents

iii

A word about units

The units used in each section of the specification are mentioned at the start of each section. Some of the units are used in all of the sections. It is important to have an overview of the unit system (SI units) and how they work.

Mechanics – Student Book Sections A, D & E

The base units for Mechanics, used in sections 1, 4 and 5 of the specification are:

Section 1 – Student Book Section A

- MASS: kilogram, **kg**
- LENGTH: metre, **m**
- TIME: second, **s**

All other units in mechanics are derived from these. Some have no special names, for example:

- VELOCITY: metre/second, **m/s**   from the definition

$$\text{velocity} = \frac{\text{distance travelled}}{\text{time taken}}, \qquad \text{hence } \frac{\text{metre}}{\text{second}}$$

- ACCELERATION: metre/second^2, **m/s^2**   from the definition

$$\text{acceleration} = \frac{\text{increase in velocity}}{\text{time taken}}, \qquad \text{hence } \frac{\text{metre}}{\text{second}} \div \text{second}$$

Others are named after famous scientists, for example:

- FORCE: newton, **N**   from the definition

$$\text{force} = \text{mass} \times \text{acceleration} \qquad \text{hence } \text{kilogram} \times \frac{\text{metre}}{\text{second}^2}$$

The name is much more convenient to use than the base units, but sometimes the units in a question will help you to remember an equation that is not given in the exam paper; this is discussed in more detail in the Exam Technique chapter.

The rest of the section 1 units are:

- Gravitational Field Strength: newton/kilogram, **N/kg**
- Momentum: kilogram metre/second, **kg m/s**

Section 4 – Student Book Section D

- WORK: joule, **J**   from the definition

  and ENERGY: work = force $\times$ distance   hence newton.metre

- POWER: watt, **W**   from the definition

$$\text{power} = \frac{\text{energy transferred}}{\text{time}}, \qquad \text{hence } \frac{\text{joule}}{\text{second}}$$

Section 5 – Student Book Section E

- VOLUME: **m³**   from the definition

  volume = length × depth × breadth

- PRESSURE: pascal, **Pa**   from the definition

  $$\text{pressure} = \frac{\text{force}}{\text{area}}$$ hence $\dfrac{\text{newton}}{\text{square metre}}$

- DENSITY: **kg/m³**   from the definition

  $$\text{density} = \frac{\text{mass}}{\text{volume}}$$

Electricity – Student Book Sections B and F

The base unit in electricity is:

- CURRENT: ampere, **A**

All other units in electricity are derived from this and the units used in mechanics:

- CHARGE: coulomb, **C**   from the definition

  charge = current × time   hence amp.second

- VOLTAGE: volt, **V**   from the definition

  $$\text{voltage} = \frac{\text{energy transferred}}{\text{coulomb}}$$ hence $\dfrac{\text{joule}}{\text{coulomb}}$

- RESISTANCE: ohm, **Ω**   from the definition

  $$\text{resistance} = \frac{\text{voltage}}{\text{current}}$$ hence $\dfrac{\text{volt}}{\text{amp}}$

Waves – Student Book Section C

- FREQUENCY: hertz, **Hz**   from the definition

  number of waves per second   hence waves/s

Thermal energy – Student Book Section D

- TEMPERATURE: kelvin, **K**

The base unit in temperature measurement in science is the kelvin, and notice the fussy little degree circle is not required. However, in daily life, temperature is usually expressed in a Celsius temperature, °C, and your laboratory thermometers are calibrated in Celsius. The size of temperature difference indicated by one division on each of these scales is the same → 1°C *change in temperature* is exactly the same as a *change* of 1K. It is important to convert to kelvin or absolute temperature when using the gas law formulae, as is explained in that section.

Radioactivity – Student Book Section G

- Rate of decay: becquerel, **Bq**

A becquerel is one decay per second.

You are also expected to be familiar with grams, centimetres, kilometres, minutes, hours, etc. and the degree, °, as a unit for measuring angles.

How to use this book

This is a revision guide and as such deals with the material in the current Edexcel specification (from 2011). It does not, therefore, have space to do justice to the subject; hopefully you are already interested in physics. The primary object is to help you to do well in your exams in physics. To do this you have to use this guide as **one** of your revision resources. You should refer back to the Student Book and ask questions in class whenever a point seems unclear. I recommend that you add additional notes to this guide to point out things you have got wrong in the past, and to add your own personal ways of remembering things – highlighter pens might be one way to do this.

The main way to prepare for the exam is to do examples, both from this book and using past papers available from Edexcel. Don't worry about making mistakes, but do learn from them! Mistakes along the way do not matter, what matters is your performance on the day of the exam. Preparation for the exam will help you to feel confident and less nervous in the exam. Spread this preparation over a sensible period of time, leaving it to the last minute can have exactly the opposite effect on your confidence!

Once in the examination room read through the questions carefully and answer questions that you feel confident about first. You must not miss easy marks through misreading questions or not getting to questions you can do before time runs out. Be familiar with the given information at the front of the paper – use it all the time when doing examples.

Some of the advice that follows may seem obvious...

Never leave questions unattempted! Even if you are making a desperate guess write it down – a blank space will never give you a mark but now and again a guess will! Some question parts will require you to use an answer from a previous part of the question; even if your previous answer is wrong, if you use it correctly in a subsequent part you can achieve full marks for that part (this is referred to as error carried forward or ecf by exam markers).

Do show your method – any question involving calculations that has more than [1] allocated to it will have method marks. A wrong answer with no method shown will get no marks, a wrong answer showing a correct method will gain 50% or more of the marks available.

Do write legibly. (If you have a diagnosed reason for having trouble with this ensure that you have any special help that you are entitled to months before the exam date.) Cross through things that are wrong, neatly, but do not obliterate them – examiners are supposed to look at all your work. Take care that numerals like 0 and 6 are distinct and that decimal points are visible to the naked eye!

Some questions will require you to include appropriate units, leaving units out will be penalised by the examiner!

Take note of the space available for answers and the number of marks allocated to a question. One line and one mark means a single word answer or a short sentence will be enough. Generally there should be as many points made in your answer as there are marks available – here is an example:

'What effect does temperature have on how well a metal wire conducts electricity?' *[2]*

Weak (one mark answer): 'It changes the resistance.'

Strong (two mark answer): 'Increasing the temperature increases the wire's resistance.'

Look for clues in the paper! Sometimes you will be asked for a definition or equation (not given at the front, of course) and later in that question or another question later on there will be a give away! Here is an example:

'Give the equation for average speed.'

[2]

Then, later in the question, there may be a speed given with its units: m/s → the units indicate speed is distance divided by time.

Be aware of exam vocabulary. Questions that say 'state' or 'describe' will be concerned with recalling a fact; the words 'explain' or 'calculate' are asking for more understanding and will usually be worth more marks. In questions involving a given formula there should be no marks allocated to repeating the formula, but if you are required to rearrange it this will usually attract a mark. Substituting correct values into an equation also attracts marks.

Be careful to ensure that you are consistent with the use of units. Here is an example:

> 'Sea water has a density of 1050 kg/m^3, what is the mass of sea water in a small tank measuring 80 cm by 50 cm by 30 cm?'

Either convert density to g/cm^3, or, and this is probably better, convert the three dimensions of the tank to m (80 cm → 0.8 m, etc). If you do not do this you might get an answer of 126 million kilograms ... which ought to alert you that something is not quite right (an *aircraft carrier* has a mass of around 'only' 30 million kilograms).

Enjoy physics and good luck with all your exams!

Equations shown thus:

$$\text{weight} = \text{mass} \times \text{gravitational field strength}$$

are not given in the exam paper.

Equations shown thus:

$$\text{Take } g = 10 \text{ m/s}^2$$

are given in the exam paper.

Chapter 1: Movement and position

Speed

$$\text{average speed} = \frac{\text{distance travelled}}{\text{time taken}}$$

You need to know this formula and be able to rearrange it to find the following two equations:

distance travelled = speed × time taken

and

$$\text{time taken} = \frac{\text{distance travelled}}{\text{speed}}$$

Units of speed: metres per second (m/s), kilometres per hour (km/h). (Also centimetres per second (cm/s) for laboratory experiments.)

Worked Example 1

If you note the mileage reading on a car odometer at the beginning and end of a journey and measure the time taken, you can calculate the average speed:

$$\text{average speed} = \frac{(23635 - 23569)}{1.5 \text{ hours}}$$

Answer: 44 km/h

You should be able to convert speeds from m/s to km/h and km/h to m/s.

Vector and scalar quantities

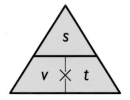

When you work out the average speed of a car, you do not need to worry about the **direction** in which the car has been travelling, just how far it has travelled. **Speed** and **distance** travelled are **scalar** quantities because they do not take direction into account. When both **size** and **direction** are important, you need to use a vector quantity. The term **displacement** *(s)* means distance travelled in a specified direction and **velocity** *(v)* means speed in a specified direction. Both are examples of **vector** quantities. You will meet further examples of scalar and vector quantities later.

Distance-time graphs

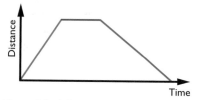

Figure 1.1 *Unless you are told otherwise distance–time graphs will show objects moving in a straight line – distance in a particular direction, i.e. displacement.*

- **Gradient** or slope gives **speed**
- **Straight** lines mean **constant** speed

The red line shows constant speed moving away from a starting point; the blue line shows that the object is stationary; the green line shows that the object has changed direction. Note that the green line has *negative gradient* (direction of travel reversed) and is *less steep* showing that the object is moving more slowly on its return journey.

Acceleration

$$\text{acceleration} = \frac{\text{change in velocity}}{\text{time taken}}$$

This formula can also be written: acceleration = (final velocity – initial velocity)/time taken

Units of acceleration: metres per second squared (m/s^2). Acceleration is a **vector** quantity.

Velocity-time graphs

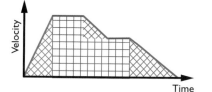

Figure 1.2 *Here velocity is used – speed in a particular direction.*

- **Gradient** or slope gives **acceleration**
- **Straight** lines mean **constant** acceleration
- **Area** under the graph gives **distance** travelled

The red line shows constant acceleration; the 1st horizontal blue line shows that the object has constant velocity; the 1st green line shows the object decelerating to a lower constant velocity (2nd blue line); the 2nd green line shows object decelerating to rest. Note: (i) steepness shows rate of acceleration or deceleration; (ii) the **area** under the graph shows the **distance travelled** during the motion.

To calculate the area under a velocity–time graph, divide the graph into rectangles and triangles, as shown in Figure 1.2, and add these areas together. Remember to use consistent units when calculating the areas. For example, if velocity is in **m/s** then time must be in **s** to give an area in **m** (the distance travelled).

Practical work

You should be able to describe experiments that allow you to measure how the distance travelled by a moving object changes with time and how you can use your results to calculate velocity and acceleration. The Student Book describes the use of ticker timers in some detail and mentions linear air tracks and electronic timing devices in other examples. You may be asked to discuss similar experiments that you may not have met in either the Student Book or during your lessons. These experiments will address the following key points:

- A **suitable** means of measuring distances travelled to an acceptable level of accuracy.
- A **sufficiently accurate** method of measuring how long it takes to travel these distances.

A simple experiment for measuring the average speed of a car involves using a hand-held stopwatch to time a car travelling over a measured distance of 50 m. A suitable means of measuring distance, in this case, would be a 'click' wheel, or a long tape measure borrowed from the PE department (but not a 30 cm ruler!). A hand-held stopwatch is *only just* acceptable as cars in a 30 mph zone will cover this distance (50 m) in 3–4 seconds. Therefore human reaction time will produce significant uncertainty in the time measurements. Having a group of people all timing (minimum of 3) will allow an **average** value to be found which will improve the overall accuracy. If the cars were travelling faster you would need to improve the accuracy of the experiment, either by having an electronic timing system, or by increasing the distance over which the car is being timed, or both.

Chapter 2: Forces and shape

Forces

Forces are pushes or pulls that one body exerts on another. You need to know that forces can change the way things are moving, make things start to move or stop moving and change the shape of things.

You should be able to name and label the forces that are acting on objects in a variety of different situations:

- **Weight**, the force that acts on a body because of gravity.

- **Friction**, the force that opposes motion, either when you try to make something move or whilst it is moving.

- **Air resistance (or drag)**, friction between an object and the air (or gas) that it is moving through.

- **Viscous drag**, similar to air resistance, but occurs when an object is moving through a liquid.

- **Upthrust**, the upward force that liquids and gases exert on objects.

- **Magnetic**, forces that magnets exert on other magnets or things made of iron (or other ferrous materials).

- **Electrostatic**, the force between electrically charged objects.

- **Normal reaction**, the special name for the **contact force** that acts on an object pressing down on another object.

- **Tension**, in strings, cables, ropes, etc. that are being stretched (not slack).

Worked Example 2

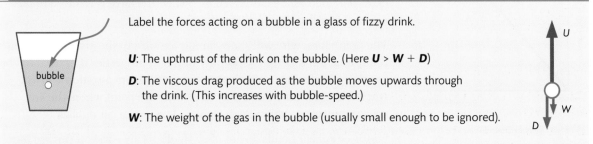

Label the forces acting on a bubble in a glass of fizzy drink.

U: The upthrust of the drink on the bubble. (Here *U* > *W* + *D*)

D: The viscous drag produced as the bubble moves upwards through the drink. (This increases with bubble-speed.)

W: The weight of the gas in the bubble (usually small enough to be ignored).

While the forces are unbalanced, as shown, the bubble will accelerate upwards.

The unit of force

Forces are measured in newtons (N). 1N is roughly the weight of an apple due to the Earth's gravity.

Balanced and unbalanced forces

Forces are **vector** quantities – the direction in which they act makes a difference to the effect that they have on the objects that they act upon. At this level you will have to deal only with forces that act in one dimension. In the previous example all the forces act in the vertical direction; to distinguish the upward forces from the downward forces we might say upward forces are positive and downward forces are negative (or the other way round, the choice is up to you). If the forces add up to zero, i.e. the total upward acting forces equal the total of the downward acting forces, the forces are said to be **balanced**. If, as in the example, the upward force is bigger than the sum of the downward forces, then the forces are said to be **unbalanced** and there is a **resultant force** acting in the up direction on the bubble.

Unbalanced forces acting on an object will cause a change in the way the object is moving:

- it will speed up (accelerate),
- slow down (decelerate) or
- change the direction in which it is moving.

Hooke's Law

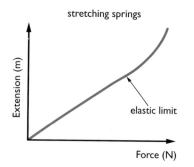

Forces can make things change shape. Robert Hooke discovered that the increase in length (extension) of a spring is directly proportional to the force pulling on it. This is shown by the *straight* part of the graph. If stretched beyond the **elastic limit**, shown where the graph starts to curve, the spring will **not** return to its original length.

You should also know how springs, wires and elastic bands behave when stretched. These are shown in the graphs below. Note that these graphs are plotted with **_force_** on the vertical axis and **_extension_** on the horizontal axis. When given an exam question on this topic be sure to note which way round the axes are labelled.

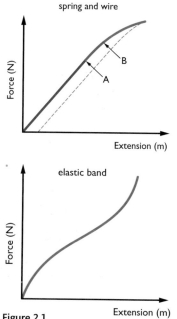

The red part of the graph shows the spring being overstretched and it will **not** return to its original length when the force is removed; the extension follows the dotted red line when the force is reduced to zero again. B is called the elastic limit.

The graph for a wire will show similar features to the graph shown for the spring, but different wires have different properties that need not be considered here. It is unlikely that any GCSE exam question will ask you about the elastic limit, only to identify the point where the line starts to curve, A, after which Hooke's Law no longer applies.

Elastic or rubber bands do not obey Hooke's Law – as you can see the graph is definitely not a straight line. The slope of the line is quite steep at the start and again at the end. If you keep on stretching the band it will break, of course. Provided the band is not stretched beyond its breaking point it remains elastic and returns to its original length when the stretching force is removed.

Figure 2.1

Practical work

The practical work in this chapter should include an experimental investigation of Hooke's Law using a spring. As with all practical work you will need to describe the measuring apparatus you use and how you take steps to make your measurements as accurate as possible. Usually the force on the spring is applied by hanging known masses on the end of the spring (100 grams has a weight of approximately 1N), and the extension is measured against the scale on a half metre rule. You will also need to take a suitable range of measurements *without* overstretching the spring, and display your results on a graph. You should know what happens when a spring *is* overstretched.

You should investigate the behaviour of an elastic band. This does not obey Hooke's Law. The extension is not proportional to the stretching force over the whole range and it remains elastic (returns to its original length) even when stretched beyond the linear part of the graph.

You should investigate or research stretching a metal wire (thin copper wire is usually used because it is not as stiff as steel), but measurement of the tiny extensions produced to an acceptable degree of accuracy is difficult without a better measuring instrument than a half metre rule.

Another experiment is an investigation of the friction force between two surfaces. This is described in the Student Book. You need to be able to measure the force needed to make the block start to slide, and to show that only one factor (say, contact force) is changed at a time. You could investigate how the friction force between two surfaces depends on:

• the contact force (normal reaction) between the two surfaces,

• the area of the surfaces in contact,

• the type of surfaces in contact – the block could be pulled first on a rough surface then a smooth surface.

You may also notice that the force needed to overcome friction to get the object moving is greater than the force needed to keep it moving – if you have ever push-started a car you will have noticed this already!

Chapter 3: Forces and movement

Force, mass and acceleration

force = mass × acceleration

Forces make things speed up, slow down, change direction or change shape. Newton investigated how a force acting upon an object may cause the object to accelerate. This important equation shows that the acceleration of an object is proportional to the force acting on it (double the force → double the acceleration) if the mass is constant. It also shows that, for a given force, the *bigger* the mass of an object the *smaller* the rate of acceleration. Rearranging this equation gives:

1. $\text{acceleration} = \dfrac{\text{force}}{\text{mass}}$

and

2. $\text{mass} = \dfrac{\text{force}}{\text{acceleration}}$

Equation 1 shows that a *large* force acting on a *small* mass will cause a big acceleration.

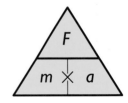

...braking

...ne brakes of a car you are relying on the friction force between the tyres and the road
...ke the car decelerate. Equation 1 on page 5 shows that this force should be as large as possible
...the deceleration as big as possible – you cannot conveniently change the mass of the car in an
...mergency! If the car skids, so the wheels are sliding across the road surface, the friction force will be
reduced; therefore over-braking has the opposite effect from what is needed. Poor tyres, water, ice or oil
on the road surface will make the maximum braking force you can apply without skidding much lower,
so your deceleration is smaller → **it takes more time and a greater distance to stop**.

Safe stopping distance

When you brake in an emergency the time it takes to stop the car and the distance the car travels before
coming to rest depend on:

- **thinking distance** – the distance you travel before you start to apply the brakes, and

- **the braking distance** – the distance you travel while the car is decelerating to rest.

The thinking distance depends on your **reaction time**. This will be longer if you are tired, if you have been
drinking alcohol or taking drugs, or if the visibility is poor.

The factors affecting braking distance are mentioned above.

Acceleration due to gravity and weight

All objects accelerate at the same rate on the Earth, provided that the effect of air resistance can be ignored.
This can be demonstrated by the 'coin and feather' experiment in which the two objects take exactly the
same time to fall through a tube with no air in it. In the exam you will be told to use the approximate value
for the acceleration due to gravity, g, at the Earth's surface – shown here.

> Take $g = 10 \text{ m/s}^2$

The force acting on an object in the Earth's gravity is called its **weight**.

> **weight = mass × gravitational field strength**

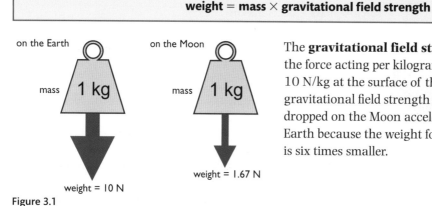

The **gravitational field strength**, g, of the Earth is
the force acting per kilogram mass and is approximately
10 N/kg at the surface of the Earth; the Moon's
gravitational field strength is only 1.67 N/kg so objects
dropped on the Moon accelerate at one sixth the rate on
Earth because the weight force acting per kilogram mass
is six times smaller.

Figure 3.1

Air resistance and terminal velocity

When objects travel through the air at a significant speed we can no longer ignore the effect of air resistance. Remember that air resistance is a force that opposes the motion of an object through the air. The size of the **drag force** caused by air resistance increases with speed – at some speed the size of the drag force will be enough to balance the weight of the object. So the overall or **resultant** force on the object will be zero so that it no longer accelerates, having reached its maximum or terminal velocity.

| Worked Example 3 |

Describe the force(s) acting on a falling object at the points labelled A, B, C and D on this velocity–time graph if the object is falling through a large distance in air.

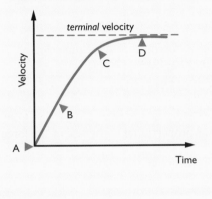

A: object at rest, **no** drag force, upthrust (**U**) remains too small to consider throughout fall, so resultant force is simply the weight (**W**) acting down.

B: object accelerating downwards, drag force (**D**) increasing but still insignificant, so resultant force is still the weight (**W**) acting down.

C: object now falling fast enough for **D** to have an effect and acts upward, opposing the movement; the resultant downward force, (**W** – **D**), is getting smaller, so the acceleration is decreasing, as is shown by the change in gradient of the line.

D: object has accelerated to the speed at which **D** balances **W** so the resultant force is zero, so acceleration is now zero and object falls with a constant velocity.

Practical work

The practical work in this chapter should include an experimental investigation of $F = ma$. Your experiment should:

- allow you to either reduce friction to a minimum (by using an air track) or compensate for friction by using a friction compensated slope,
- have a way of applying a constant force, **F**, to the object being accelerated,
- have a way of measuring this constant force,
- allow the acceleration, **a**, to be calculated.

Using a suitable instrument you should measure the mass, **m**, of the object you are accelerating and then repeat the experiment with different values of **m** and **F**.

You should be able to present your results in table and graph form, and to show from your graph that the formula is valid.

You should be able to describe and analyse experiments to measure the acceleration due to gravity. Methods for this include the use of a ticker timer or analysis of a photo of a falling object taken with multiple images produced by a stroboscope flashing at known intervals. These two methods are discussed in the Student Book. In an exam you may be given the necessary equation to analyse an electronic timing method. Electronic timing devices are necessary when the timed intervals are short and the use of human reaction time alone would produce unacceptably large errors.

Terminal velocity is often demonstrated by dropping steel ball bearings into a tall measuring cylinder filled with a viscous (thick) liquid such as glycerol. Falling winged seeds, like sycamore seeds, also reach a terminal velocity though for more complex reasons, but this is suitable as a qualitative experiment.

Chapter 4: Momentum

Momentum

$$\text{momentum} = \text{mass} \times \text{velocity}$$

Momentum is a measure of how easily an object may be brought to rest. An object with a large momentum will need a bigger force (acting for a longer time) to bring it to rest.

Momentum is a **vector** quantity, its direction is important and must be stated.

The unit of momentum

Momentum is measured in kilogram metres per second (kg m/s) provided that mass is in kg and velocity is in m/s.

Newton's 2nd law

Newton showed that the rate of change of momentum of an object was proportional to the size of the force acting on the object. Using SI units this leads to the equation:

$$\text{Force} = \frac{\text{change in momentum}}{\text{time taken}}$$

Which can also be written:

$$\text{Force} = \frac{\text{final momentum} - \text{initial momentum}}{\text{time taken}}$$

Usually the mass does not change so this simplifies to: $F = ma$ (see Chapter 3).

Conservation of momentum

The momentum of bodies colliding (or springing apart as in an explosion) is always conserved (i.e. the same before and after the event), provided that the only forces acting on the bodies are the force of one body on the other and vice versa. In practice, other forces have to be taken into account; for example friction, air resistance and weight.

If you eliminate or minimise these then you can show, by experiment, that when two bodies collide:

total momentum of the two bodies before collision = total momentum of the two bodies after collision

As momentum is a vector quantity you must take the direction of movement into account; if one object is moving north, and the other is moving south, the *sign* of one must be negative. It is up to you to choose whether to make north or south positive but once you have chosen you *must* stick to that decision.

Elastic and inelastic collisions

An **elastic** collision is one in which no energy is 'lost'. Most collisions will involve some of the movement energy being 'lost' as it is converted into heat and sound – such collisions are ***partially* elastic**. When the colliding objects do not rebound at all – so stick together – the collision is called **inelastic**.

A small pellet of mass 0.01 kg travelling at 30 m/s is fired into a block of plasticine on a small trolley which is stationary on a very low friction track (so that friction is so small we do not need to consider it). The pellet collides inelastically with the plasticine (it gets embedded in the plasticine) and the pellet, plasticine and truck move as one after the collision. If the truck and trolley have a mass of 0.19 kg, what is the speed of the trolley just after the collision?

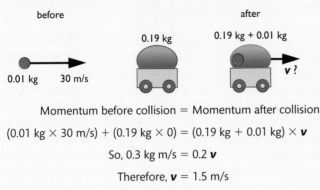

before

0.19 kg

0.01 kg   30 m/s

after

0.19 kg + 0.01 kg

v ?

Momentum before collision = Momentum after collision

$$(0.01 \text{ kg} \times 30 \text{ m/s}) + (0.19 \text{ kg} \times 0) = (0.19 \text{ kg} + 0.01 \text{ kg}) \times v$$

So, 0.3 kg m/s = 0.2 v

Therefore, v = 1.5 m/s

Car safety

Rearranging the momentum equation gives: $F \times t = mv - mu$

or, in words:

Force $\times$ time = change in momentum

When a car stops the change in momentum is fixed by the speed you are travelling at and the mass of the car + contents. If the car is brought to rest rapidly, so that t is very short, the force acting on the car *and its contents (people)* will be *very* large. The human body cannot withstand large forces without damage so, to reduce the size of the force acting during deceleration, it is vital to increase the time taken to bring the car to rest. Crumple zones increase the time it takes for the car to come to rest.

Newton's laws of motion

1. Unless a resultant force acts on an object it will continue to move in a straight line at constant speed or remain at rest.

Figure 4.1

2. The rate of change of momentum of a body is proportional to the force acting on it.

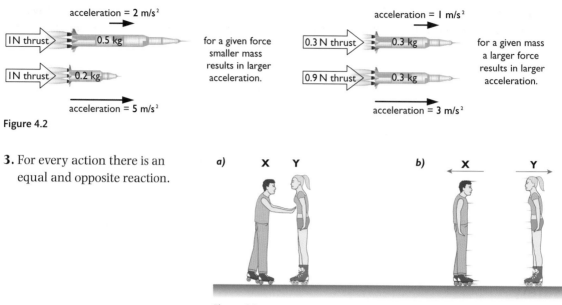

acceleration = 2 m/s²

IN thrust | 0.5 kg

IN thrust | 0.2 kg

acceleration = 5 m/s²

for a given force smaller mass results in larger acceleration.

acceleration = 1 m/s²

0.3 N thrust | 0.3 kg

0.9 N thrust | 0.3 kg

acceleration = 3 m/s²

for a given mass a larger force results in larger acceleration.

Figure 4.2

3. For every action there is an equal and opposite reaction.

a)   X   Y   b)   X   Y

Figure 4.3

Practical work

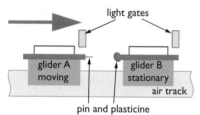

light gates

glider A moving | glider B stationary

air track

pin and plasticine

Figure 4.4

You should be able to describe an experiment to demonstrate conservation of momentum. On a *levelled* linear airtrack you should make one glider collide with another – it is simplest if the second glider is stationary. You measure the mass of both gliders. If you make the collision inelastic (so the gliders stick together) it is easier to measure the momentum after the collision. Measure the velocity of the moving glider before the collision and the velocity of the gliders after the collision has taken place.

The experiment should be repeated with different masses and velocities.

Chapter 5: The turning effect of forces

The moment of a force

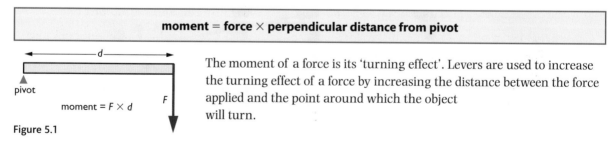

moment = force × perpendicular distance from pivot

pivot

moment = $F \times d$

F

d

Figure 5.1

The moment of a force is its 'turning effect'. Levers are used to increase the turning effect of a force by increasing the distance between the force applied and the point around which the object will turn.

In balance

Often several forces will act on an object; some may have a clockwise turning effect and others may have an anticlockwise turning effect. If the effects cancel one another out then the object will be in balance. The condition for balance is:

> ↻ sum of clockwise moments = sum of anticlockwise moments ↺

Worked Example 5

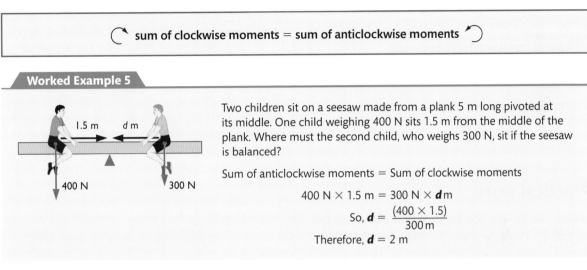

Two children sit on a seesaw made from a plank 5 m long pivoted at its middle. One child weighing 400 N sits 1.5 m from the middle of the plank. Where must the second child, who weighs 300 N, sit if the seesaw is balanced?

Sum of anticlockwise moments = Sum of clockwise moments

$$400 \text{ N} \times 1.5 \text{ m} = 300 \text{ N} \times \boldsymbol{d} \text{ m}$$

$$\text{So, } \boldsymbol{d} = \frac{(400 \times 1.5)}{300 \text{ m}}$$

$$\text{Therefore, } \boldsymbol{d} = 2 \text{ m}$$

Centre of gravity

The whole of the weight of an object acts through **one point** of the object called its **centre of gravity**. If a shape is suspended from a point, so that it can turn, it will come to rest with the centre of gravity of the object immediately below the point from which it is suspended:

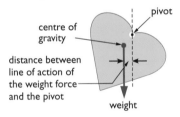

On the left the weight force is not vertically below the pivot point and so produces an anticlockwise turning effect about the pivot. This makes the shape turn and swing from side to side until the centre of gravity is immediately below the pivot as shown on the right; in this position the weight force acts along the line through the pivot so the moment of the force is zero.

Figure 5.2

Figure 5.3

Forces on a beam

If a beam is stationary two conditions must be fulfilled:

- the sum of the upward forces acting on the beam is equal to the sum of the downward acting forces,

- the sum of clockwise moments acting on the beam about any point is equal to the sum of anticlockwise moments acting on the beam.

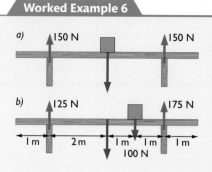

a)

150 N   150 N

A beam 6 m long weighing 200 N is supported by two posts placed 1 m from each end. A box weighing 100 N is placed a) in the middle of the beam, then b) 2 m from one end. Show the forces acting **on the beam** in each case, showing any calculations necessary.

b)

125 N   175 N

1 m   2 m   1 m 1 m   1 m
              100 N

In both cases the up and down forces on the beam must be equal, so in a) 150 N + 150 N = 200 N + 100 N and in b) 125 N + 175 N = 200 N + 100 N. In both cases the sum of the moments trying to rotate the beam about **any** point clockwise and anticlockwise must be equal. For example, take moments about the **middle** of the beam:

in a) 150 N × 2 m = 150 N × 2 m; in b) 125 N × 2 m + 100 N × 1 m = 175 N × 2 m

You will not be asked to calculate the forces in situations like b) without some guidance but you should understand that the **position** of the box affects the size of the upward forces in each supporting post.

Practical work

You should be able to show how both moments and vertical forces balance for a beam. A simple experiment supporting beams of known weight (weigh with a newton meter or multiply mass by gravitational field strength) with the supports placed as shown below will allow you to do this:

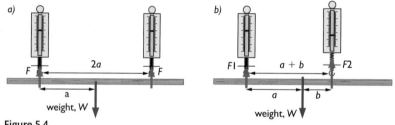

a)

F $2a$ F

a

weight, W

b)

$F1$ $a + b$ $F2$

a b

weight, W

The newton meters are held in clamps positioned to keep the beam level. The beam is supported by wire loops from the newton meters. It is assumed that the beam is uniform and therefore its centre of gravity is in the middle of the beam.

Figure 5.4

Considering the vertical forces:

In a) the beam is supported at two points that are the same distance from the middle of the bar.

So, $W = F + F$

In b) the beam is supported at two points that are at different distances from the middle of the bar.

So, $W = F1 + F2$

Considering the moments, clockwise and anticlockwise about the left hand support point:

For a): $W \times a = F \times 2a$

For b): $W \times a = F2 \times (a + b)$

You could experiment further by placing a known load at various positions along the beam.

Chapter 6: Astronomy

The solar system

You should know that our solar system consists of **planets** orbiting the Sun and that some planets also have **moons** which, in turn, orbit their planets in the same way as the planets orbit the Sun. Planets and moons are visible because they reflect light from the Sun; the Sun is a **star**, producing vast amounts of energy, like light and heat, fuelled by a continuous nuclear reaction.

Comets also orbit our Sun. They are relatively small objects consisting of rock and ice and have very eccentric orbits, that is, their paths come close to the Sun for part of the path but then travel a long way from the Sun, to the outer reaches of the solar system. They are only visible for a short period of time in their orbits, and have a 'tail' of melting ice and debris pointing away from the Sun.

Beyond our own solar system the only visible objects are stars (and huge clusters of billions of stars called **galaxies**). Our galaxy, called the Milky Way, is one of billions of galaxies that make up the **universe**.

Gravitational forces

Gravity is the force that keeps the Moon in orbit around the Earth and the moons of other planets in our solar system in their orbits. It also keeps the planets in orbit around the Sun and the Solar System in orbit around the centre of our own galaxy, the Milky Way.

> Please see the Appendix for some additional material regarding gravity.

Earth and Moon system

Figure 6.1 *Earth and Moon system*

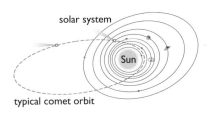

Figure 6.2 *Typical comet orbit*

Figure 6.3 *A galaxy*

The strength of gravity

Gravity is the force that attracts objects with mass to each other. Newton showed that the size of the gravitational force between objects depends on:

- the size of the masses involved. Large masses, like stars and planets, produce large gravitational forces. Smaller bodies like the Moon have much smaller gravitational forces.

- the distance between the masses. The greater the distance between two masses the smaller the gravitational force between them.

The effect of gravity between two small masses is extremely small.

Orbital speeds of satellites

Speed is the distance travelled divided by the time taken to travel the distance. The orbits of planets, moons and satellites are nearly circular, so the distance travelled is $2\pi r$ where r is the radius of the orbit. The time taken, T, is the **period** of the orbit, the time for one complete circuit. Thus orbital speed is given by:

$$\text{Orbital speed} = \frac{2 \times \pi \times r}{T}$$

Worked Example 7

The radius of the Earth's orbit around the Sun is about 150 million kilometres. The period of the Earth's orbit is one year. Calculate the average speed of the Earth as it moves around the Sun, a) in km/h, and b) in m/s.

a) Orbital speed $= \dfrac{2 \times \pi \times 150\,000\,000\,\text{km}}{365 \times 24\,\text{h}}$

So the average orbital speed is 108 000 km/h (to 3 significant figures).

b) Orbital speed $= \dfrac{2 \times \pi \times 150\,000\,000 \times \mathbf{1000}\,\text{m}}{365 \times 24 \times \mathbf{3600}\,\text{s}}$

So the average orbital speed is 30 000 m/s (to 2 significant figures).

This is a given formula; you need to be careful to use the correct units. To convert kilometres to metres multiply by 1000; to convert hours to seconds multiply by 3600 (1 h = 60 min = 60 × 60 s).

Warm-up questions

1. Mariam sprints 100 m in 12.9 s. What is her average speed in **a)** m/s, **b)** km/s and **c)** km/h?
   [Hints: 100 m = 0.1 km, 1 hour = 3600 seconds.]

2. The Sun is about 150 000 000 km distant from the Earth. Light travels at 300 000 km/s. How long does it take light from the Sun to reach the Earth in **a)** seconds and **b)** minutes?

3. The next nearest star to the Earth, Alpha Centauri, is 4.2 light years away. A light year is the distance travelled by light in one year. Use the value for the speed of light in question 2 to calculate the distance to Alpha Centauri in kilometres.

4. Look at the following distance–time graphs for moving objects:

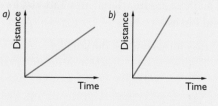

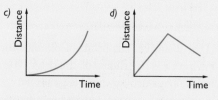

   Which graph shows an object that is
   a) Travelling at the highest *constant* speed,
   b) Changing the direction in which it is travelling,
   c) Getting faster, and
   d) Travelling at the lowest *constant* speed?

5. A car accelerates from 0 to 90 km/h in 8 s.
   a) Convert 90 km/h to m/s.
   b) Now calculate the rate of acceleration in m/s².

6. Look at the velocity-time graph below:

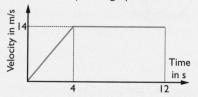

   a) Calculate the rate of acceleration during the first 4 s of the motion.
   b) Describe how the object is travelling during the period between 4 and 12 seconds.
   c) How far does the object travel during the 12 seconds shown?
   d) What is the average velocity of the object during the first 4 seconds?

7. A boat is floating on a pond.
   a) Draw a labelled diagram showing the *two* forces that act on the boat.
   b) These two forces are *balanced*. Explain what this means.

8. A helium balloon is tied with string to the ground on a still day.
   a) Draw a labelled diagram showing the *three* forces that act on the balloon.
   b) How will your force diagram change if the string is cut, and
   c) What effect will this have on the balloon?
   d) After the string has been cut a new force will start to act on the balloon; what is the name of this force and what effect will it have on the behaviour of the balloon?

9 State the *three* effects that an unbalanced (resultant) force may have on the way an object moves.

10 Look at the graph below which shows how the extension of a spring varies with the force pulling on it:

a) At which point on the graph does the spring stop obeying Hooke's Law?

b) What is the *name* of the point on the graph where the spring stops obeying Hooke's Law?

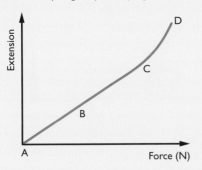

11 a) What *resultant* force must act on a ship with a mass of 30 000 000 kg to make it accelerate at 1.8 m/s²?

b) What force will oppose the acceleration?

c) What does this mean about the size of the thrust force that must be provided by the ship's propellers?

12

The upward thrust provided by the rocket engines of the lunar module (LEM) on take-off was 16 000 N; the weight of the LEM on take-off was 7500 N. Assume that the Moon's gravity is one-sixth of that on Earth.

a) What was the resultant upward force on the LEM?

b) What, therefore, was the initial acceleration of the LEM?

[Harder:]

c) Why will the acceleration increase for a while after take-off?

13 A rifle bullet is pushed with an average force of 800 N causing it to accelerate at 100 000 m/s². What is the mass of the bullet?

14 The following graphs show how the velocities of 3 cars change with time in the event of an emergency stop (assume the scales on all three graphs are the same):

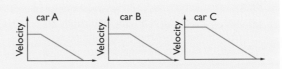

a) Which car travels the greatest distance before coming to rest? How can you tell this from the graphs?

b) Which car's driver reacted most quickly to the emergency? How can you tell this from the graphs? Give two reasons that could account for the slower reactions in the other two cars.

c) Use the graphs to justify the advice to slow down in conditions of poor visibility.

d) Which part of the graphs would change for *all 3 cars* if the road surface became more slippery?

15 The Apollo 15 astronaut, Commander David Scott, repeated Gallileo's famous experiment on the Moon, by dropping a feather and a hammer at the same time.

a) What was the outcome of this experiment and what did it show?

b) If the experiment was repeated with the same feather and hammer on the Earth the outcome would be different for *two* reasons – what are they and how would they affect the way in which the feather and hammer behaved?

16 The graph below shows an object that has fallen from a considerable height through the air:

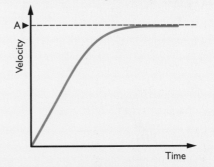

a) The first part of this velocity time graph is practically a straight line; what does this tell us about the acceleration of the object and, therefore, the resultant force on the object?

b) Why does the graph start to curve?

c) The acceleration of the falling object becomes zero at the value of velocity, **A**; what is this value of velocity called?

d) When the acceleration has become zero what is the resultant force on the object?

17 An astronaut of mass 100 kg is working on a space station. She is not moving relative to the space station. She throws an equipment box of mass 2.5 kg away from herself at a velocity of 12 m/s.

   a) What is the momentum of the equipment box?

   b) What is the momentum of the astronaut after throwing the box?

   c) What is her velocity after throwing the box?

18 A railway truck of mass 4000 kg rolls into a stationary truck of mass 6000 kg at a velocity of 8 m/s. After the collision the trucks are joined together. What is the velocity of the two trucks immediately after the collision?

19 *a)* A test car of mass 600 kg travelling at 12 m/s strikes a wall and is brought to rest in 0.02 s. What is the average force acting on the car during the collision?

   b) A different make of car undergoes the same test and takes 0.25 s to come to rest. Why is this a safer car?

20 Newton found that 'every action has an equal and opposite reaction'. Explain what this means using the example of the astronaut in question 17.

21 Put the following forces in order of their turning moments about the pivots shown in ascending order:

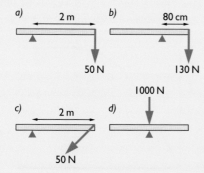

22 *a)* Put the following objects in order from the one with the lowest centre of gravity to the one with the highest centre of gravity.

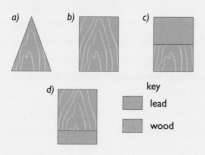

   b) Which is **(i)** the most stable object and **(ii)** the least stable (most likely to topple)?

23 What is the size of the missing force in each of the following situations showing a beam in equilibrium? Each beam is of uniform density.

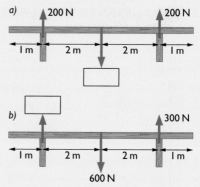

24 The Moon and the Sun both shine light down onto the Earth. How is the light we receive from these two bodies different?

25 *a)* What force keeps the Moon in orbit around the Earth and the Earth in orbit around the Sun?

   b) Identical satellites are placed in orbit around three planets:

   Planets **A** and **C** have the same mass; planet **B** is twice as dense. The satellite orbiting planet **C** has an orbit of twice the diameter of the other two. Put the satellites in order of the size of the force of gravity they feel, in ascending order.

26 An astronaut has a one kilogram mass and two spring balances, one calibrated to measure forces in newtons the other calibrated to measure mass in kilograms. The astronaut hangs the 1 kg from both instruments; what would you expect the two balances to read *a)* on the Earth's surface and *b)* on the Moon's surface?

27 Copy and complete the following sentences about our universe, filling in the spaces with the correct words:

Our solar system is made up of a _____ called the _____ which is orbited by a number of _____. Some of these planets have _____ orbiting them, including the Earth. There are billions of _____ in our _____ which is called the Milky Way. The universe itself is made up billions of _____.

28 An astronomer uses a medium power telescope to study some objects in the sky. He sees the following:

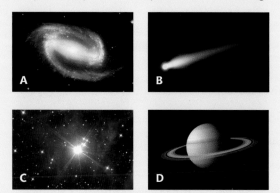

Select the correct set of labels for these objects:

1: **A** is a comet, **B** is a planet, **C** is a star, **D** is a galaxy

2: **A** is a comet, **B** is a planet, **C** is a galaxy, **D** is a star

3: **A** is a planet, **B** is a comet, **C** is a star, **D** is galaxy

4: **A** is a galaxy, **B** is a comet, **C** is a star, **D** is a planet

29 The diagram below shows two objects orbiting around the Sun. Which of the following statements is true?

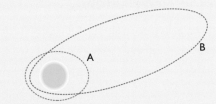

1: **A** is the orbit of a comet, **B** is the orbit of a planet

2: **A** is the orbit of a planet, **B** is the orbit of a comet

3: Both show orbits of a comet

30 The formula for calculating the orbital speed, v, of a planet around a star is
$$v = \frac{2 \times \pi \times r}{T}$$
Use this formula to calculate the orbital speed, in m/s *and* km/h, of the Earth which has an orbital period of 365 days and radius 150 million kilometres, and Mercury which has an orbital period of 90 days and radius 60 million kilometres.

Section A: Forces and Motion

Section B: Electricity

Chapter 7: Mains electricity

Using electricity safely

You should be able to identify the following hazards (shown in Figure 7.1) that increase the chances of severe and possibly fatal electric shocks:

- frayed cables, any damaged insulation can expose 'live' wires,

- long cables, as they are more likely to get damaged or trip people up,

- damage to plugs or any insulating casing on any mains operated devices,

- water around electric sockets or mains operated devices,

- pushing metal (conducting) objects into mains sockets – usually only a problem with very young children, solved by using socket covers.

Figure 7.1

You should also know how to wire a three pin plug correctly with the different coloured wires in the correct places, as shown here.

The cable insulation must be firmly gripped by the cable anchor and the correct fuse fitted, according to the type of appliance to which the plug is connected.

In addition you should be aware of the following safety features:

- Insulation – all mains wiring is double insulated with two layers of insulation. This prevents the separate conductors (live, neutral and earth) from touching and prevents anyone from touching a 'live' (mains voltage) wire.

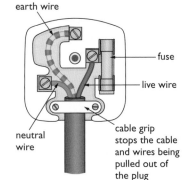

Figure 7.2

- Some appliances are **double insulated**; as well as all their wiring being insulated the outer casing of the appliance is also made of an insulating material, usually plastic. This means there is no chance of an electric shock from the casing – double insulation is often used with electric kettles and power tools like electric drills.

- Appliances with a metal outer casing that the user might touch must be **earthed**. The earth wire ensures that the outer casing is held at 0 V and provides a very low resistance path for current in the event of a fault in which the live wire touches the casing. This means that the fault current will be very large and cause the fuse to blow immediately, disconnecting the live supply.

- **Fuses** are fitted in plugs, and in the consumer unit next to the electricity meter. Cartridge fuses are ceramic, heat proof tubes containing a wire designed to melt when a specified current size is exceeded, thereby cutting off the live supply. Fuses have to be replaced when they have blown.

- **Circuit breakers** are now used in domestic consumer units rather than fuse wire or cartridge fuses. They perform the same job as fuses, breaking the live connection when a specified current size is exceeded. Circuit breakers are usually operated magnetically and may be reset by simply pressing a button.

It is important to be aware that RCCBs (residual current circuit breakers) differ from the individual circuit breakers used in consumer units – they are designed to disconnect the live supply ('trip') at much lower current levels. These devices are *essential* when using electric appliances in hazardous conditions, such as mowing a damp lawn where the danger of a fatal electric shock is much greater.

An obvious safety feature used in mains electric circuits is the switch. Most switches break just one wire in the circuit (single pole switches) and this **must be the live wire**. Some appliances, notably electric cookers, showers and immersion heaters break both the live and neutral connections (double pole switches) for extra safety.

Figure 7.3 *Inside the consumer unit*

Figure 7.4 *RCCB for individual appliance*

The heating effect of an electric current

When an electric current is passing through a wire some of the electrical energy is converted into other forms; some useful, some not. A common *useful* conversion is electrical energy into heat. This is widely used in our homes in electric irons, cookers, kettles, hairdriers, etc. The cables that carry electricity to and around our homes have very low electrical resistance in order to make the unwanted conversion of electricity to heat as small as possible. It is important to realise that these cables still have some resistance and, if the current flowing in them is too big, the cables themselves will overheat and may even cause a fire. It is for this reason that fuses are used – to stop cables overheating – and why it is essential to use the correct fuse (or circuit breaker) for each circuit or individual appliance.

Electrical power

You should be able to recall the following equation that allows you to calculate the **power**, in **watts**, produced in an electric appliance when a **current**, in **amps**, is passed through it by a **voltage**, in **volts**.

power = current × voltage

You can use the triangle method to rearrange the formula to calculate current or voltage.

$$\begin{array}{c} P \\ \hline I \times V \end{array}$$

Worked Example 1

You have an electric drill with a power rating of 750 W designed to run from the 230 V mains supply. Use this information to calculate the current it will pass when working normally.

Rearrange the formula given above to get current = $\dfrac{\text{power}}{\text{voltage}}$

then substitute the values given, so current = $\dfrac{750\,\text{W}}{230\,\text{V}}$

Therefore, the normal working current for this appliance is 3.26 A

Power is the rate at which electrical energy is converted in an appliance; in the above example the drill is converting 750 joules of energy per second into other forms (mainly useful movement energy but also unwanted sound and heat energy). If you wish to calculate how much energy an electrical appliance uses in a given period of time you should use:

$$\text{energy} = \text{power} \times \text{time}$$

which, when you substitute power = current × voltage, becomes the *given* equation:

$$\text{energy} = \text{current} \times \text{voltage} \times \text{time}$$

You must use the correct units; **energy** is measured in **joules**, provided you have used **amps** for the current, **volts** for the voltage and **seconds** for the time.

Worked Example 2

An electric kettle draws a current of 10 A when connected to a 230 V mains supply. Calculate how much energy it uses, in joules, if it takes 3 minutes to boil some water.

Here you can use the given equation: energy = current × voltage × time *but* you must convert time in minutes to time in seconds. So,

energy = 10 A × 230 V × (3 × 60 s)

Therefore, the energy used in this case is 414 000 J (or 414 kJ, but note that *this* question asks for the energy used **in joules** not kilojoules).

Alternating current and direct current (AC and DC)

You should be able to understand the difference between alternating electrical power supplies, like the mains supply in this country, and direct supplies like batteries. A battery makes electricity flow in one direction only, this is called a **direct current** (DC). An alternating supply, like the mains, causes the current to change continuously, with electricity flowing in one direction then the other; this is **alternating current** (AC).

Practical work

The majority of the practical work for this section is covered in later chapters. You should be able to wire a three pin plug safely and select an appropriate fuse for the appliance to which it is connected.

Chapter 8: Electric charge

Conductors and insulators

Materials that conduct electricity well are known as **conductors**. They are usually metallic; copper, silver and gold are all very good conductors of electricity. Materials that do not conduct electricity at all are called **insulators**. These are usually non-metallic; rubber, glass and many types of plastic do not conduct electricity. Cut wood, when dry, may also be considered to be an insulator.

Charges within an atom

This chapter is about static electricity – where objects have an overall electric charge and there is no path for it to *move* from or through the object. It is important that you are aware of the basic structure of the atom in order to understand how objects can be given a static charge.

The atom is made up of **neutrons** and **protons,** which together make up the nucleus of the atom. The nucleus is surrounded by fast moving **electrons**.

The key points to remember are:

- Protons are **positively** charged.

- Neutrons carry **no electric charge**, they are neutral particles.

- Electrons are **negatively** charged.

- The charges on the proton and electron are **equal in size**, but of opposite sign.

- All atoms have equal numbers of electrons and protons, so the charges balance out and atoms are, overall, **neutral**.

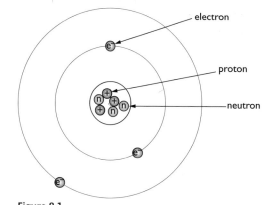

Figure 8.1

In solids the positions of atoms are fixed and therefore the nuclei and the protons within them are fixed; the surrounding electrons can be made to move from place to place under certain conditions and it is the **transfer of electrons from place to place** which is key to understanding how both static and current (moving) electricity occur.

Charging objects by friction

The first observations of static electricity were made in Greece more than 2500 years ago. Rubbing an insulating material called amber with fur made it attract light objects like feathers – you can repeat this experiment by rubbing a clear plastic ruler or plastic comb with a dry cloth and attracting small pieces of dry paper to it. It was also noticed that different materials became charged in one of two ways, as may be demonstrated by the following experiments.

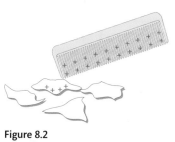

Figure 8.2

Polythene and clear acetate rods can be easily charged (on a dry day) by rubbing them with a dry cloth. Both can be shown to be charged by demonstrating that they produce repulsion as shown in Figure 8.3a and b. The final part of the experiment, Figure 8.3c, which shows that two objects that are charged may attract each other, shows that the rods must carry two distinct and different types of charge.

the negatively charged polythene rods **repel**; so do the positively charged acetate strips.

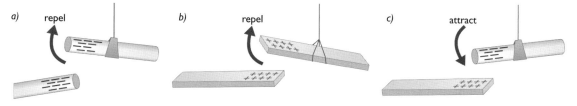

a) repel   b) repel   c) attract

objects with *unlike* charges **attract**

Figure 8.3

The two different types of charge were named **positive** and **negative** charge and the simple rule:

> **unlike charges attract, like charges repel**

is clearly demonstrated by the experiment.

The process of charging by friction is explained as follows:

> When rubbing two materials together some electrons are torn from the surface of one of the materials and transferred to the other. The material which **loses electrons** now **has a net positive charge** because not all the positive charges in the atomic nuclei are balanced by the negative charges of the electrons. The material that **gained electrons** now has more negative charge than positive charge and is therefore **negatively charged**.

It is possible to demonstrate that the cloth and rod used in this type of experiment will always be equally and oppositely charged since one gains exactly the same amount of charge (number of electrons) as the other loses, but these experiments are difficult to do with school apparatus.

Forces between charged and uncharged objects

If you rub a balloon vigorously on a dry jumper it will become charged and you may be able to 'stick' it to a wall. The wall surface is likely to be insulating and will also be uncharged (all protons exactly balanced by the same number of electrons). You can show that there is always attraction between a charged object and an uncharged object.

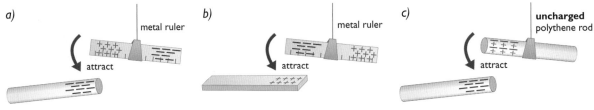

Figure 8.4

The explanation of the behaviour in the first and second figures is relatively straightforward. In metals, electrons can move freely throughout the whole of the metal object, so an induced charge of the opposite sign is always produced nearest to the charging object. In the third figure the electrons are bound to the vicinity of their particular nucleus, but the way the charge is distributed around the atom is affected, with one side of each atom being slightly more positive than the other. You will not be expected to explain this in detail at GCSE.

Uses of static electricity

You should be able to describe how the following devices work:

- INKJET PRINTERS: charging ink droplets in inkjet printers allows the droplets to be directed to particular places on the paper by deflecting them between charged plates.

- PHOTOCOPIERS: A statically charged drum is exposed to light, reflected from the document to be copied, which discharges the drum everywhere except where the dark print does not reflect light. The charged parts of the drum attract the toner which is then transferred to the printing paper. Heat then bonds the toner particles to the printing paper.

- PAINT SPRAYING: The tiny droplets of paint are given a static charge and the object to be painted is connected to a supply of opposite charge. This causes the paint droplets to be attracted to the object being painted and the amount of paint wasted is drastically reduced and a more even finish is produced.

- ELECTROSTATIC PRECIPITATORS: The small particles of soot and other dust produced in burning waste materials are given a static charge and are then passed through a highly charged grid which attracts the dust particles, stopping them from escaping into the atmosphere.

Problems with static electricity

- ELECTRIC SHOCKS: Cars become charged with static electricity, particularly on dry days, and can give an unpleasant shock when someone touches the car. This also happens while walking on acrylic carpets.

- FUELLING TANKERS and AIRCRAFT: When fuelling it is possible for static charge to build up on planes or tankers and, should a spark occur, a fire or explosion could result. This is prevented by ensuring that the tanker or plane is electrically earthed to discharge them.

- HANDLING MICROPROCESSORS and COMPUTER 'CHIPS': Workers handling electronic components must take care not to become charged by static as this can easily destroy expensive components. They wear earthing straps and work on earthed metal benches to prevent this.

Practical work

Apart from the simple experiments described above there is little practical work in this chapter. It is, however, worth pointing out that experiments with static electricity are more likely to work on dry days and that all apparatus should be thoroughly dried.

The amount of electric charge involved in static electricity experiments is tiny, but even quite small amounts of charge building up on an object can produce large voltages – it takes around 10 000 V to make a spark jump across an air gap of 1 cm so even tiny sparks of a millimetre require large voltages. If you receive a static shock, the small amount of electricity involved means that you are very unlikely to suffer any long-term effects!

Chapter 9: Current and voltage in circuits

Conductors, insulators and electric current

Metals are good conductors of electricity because they have large numbers of **'free' electrons**; these are electrons that are not bound to any particular atom in the structure of the metal, they are free to move at random throughout the metal. When a voltage is applied across a metal wire, an electric force acts on the electrons causing them to 'drift' in the direction of this electric force. As the electrons carry **negative electric charge** this results in a movement of charge along the wire:

> electric current is the rate of flow of electric charge through a conductor

Electric charge is measured in coulombs (unit symbol, C). The amount of negative charge on an electron is really tiny; **6.25 *trillion*** electrons have a total charge of **−1 C**. (I have taken one trillion = 1 000 000 000 000 000 000 that is, 10^{18}, but common usage would be to say one billion billion.)

This leads to a simple equation for current:

$$\text{current in amps} = \frac{\text{charge in coulombs}}{\text{time in seconds}}$$

A current of 1 A means that charge is passing through a conductor at a rate of 1 C/s

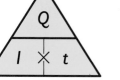

This is usually rearranged as: charge, **Q** in coulombs = current, **I** in amps × time, **t** in seconds

A small torch lamp passes a current of 150 mA when switched on. Calculate how many coulombs of charge pass through the lamp if the torch is turned on for 15 minutes.

Write the appropriate form of the equation: charge = current × time **but** to obtain the amount of charge in *coulombs* you must convert time in minutes to time in *seconds* and current in milliamps to current in *amps*.

So, charge = 0.15 A × (15 × 60) s

Therefore, the amount of charge that has passed through the lamp is 135 C

Voltage

Batteries, dynamos, photovoltaic cells are all ways of providing the energy need to make electric charge move around a circuit, that is, to make a current flow. The 'strength' of these energy sources is measured in **volts**, and the voltage of an energy source like a battery tells us how much energy is transferred per coulomb of charge that is passed through the battery:

1 volt means 1 joule per coulomb

So a 12 V car battery supplies 12 J of energy to each coulomb of charge that it circulates.

This energy is converted to other forms as the charge flows through the components that make up the circuit that the battery is supplying.

Measuring current and voltage

Current is measured using an **ammeter** placed in **series** with the component through which the current passes, as shown in Figure 9.1.

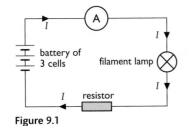

Figure 9.1

The current, *I*, is the same throughout the circuit. Since an ammeter is a part of the circuit it is important that it has very little **resistance**, otherwise it will make the current in the circuit smaller. Remember the ammeter tells you how many coulombs of charge are passing through the lamp per second.

Figure 9.2 *Ammeters*

Voltage is measured using a **voltmeter** placed in **parallel** with the component through which the current is flowing, as shown in Figure 9.3.

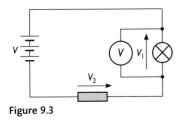

Figure 9.3

The voltage, V_1, across the lamp tells you how many joules of energy are being converted to heat and light per coulomb of charge passing through the lamp. **Note that** $V_1 + V_2 = V$, the battery voltage.

Voltmeters have very large resistance.

Figure 9.4 *Voltmeters*

Electrical circuit symbols

Circuit symbols are simple representations of components designed to make drawing circuit diagrams quicker and clearer. You should be able to recognise and draw the following component symbols as well as those already introduced in this chapter (see page 242 of Student Book).

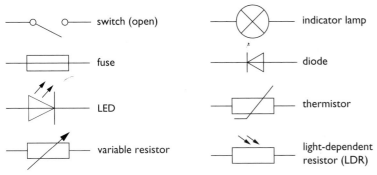

switch (open)

indicator lamp

fuse

diode

LED

thermistor

variable resistor

light-dependent resistor (LDR)

Figure 9.5

Series and parallel circuits

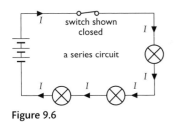

Figure 9.6

You have already seen a series circuit in this chapter. In a **series** circuit the same current flows through all the components as there are no alternative routes at any point.

Any break in the circuit, either if the switch is opened or one of the lamps 'blows' (the filament melts or breaks), will stop the current flowing in the circuit and *all the lamps* will turn off.

If the lamps are identical, each will get an equal share of the battery voltage and they will all have the same brightness.

A **parallel** circuit will contain points where the current can split to take two or more routes. These points are shown on Figure 9.7 by the join 'blobs'. At the first join the current, I, divides into I_1 flowing through lamp 1 and I_2 flowing through lamp 2. **Note that:** $I = I_1 + I_2$

At the second join the two currents combine and the original current, I, returns to the battery.

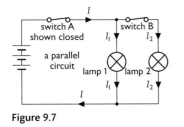

Figure 9.7

If the lamps are identical the current will divide equally, and since both have the same (battery) voltage across them they will light with equal brightness. If lamp 2 blows or if switch B is opened, lamp 1 will be unaffected. However, the size of current, I, supplied by the battery will halve. Note that current, I, will not divide equally if lamps 1 and 2 are not identical.

Decorative lights used on festive occasions are usually wired in series. Fifty lights in series means that each light has 1/50 of 230 V connected across it – so each light will have a working voltage of 4.6 V. This means that replacing a bulb is safer, but all the bulbs go out if an individual light fails.

Mains socket outlets and mains lights in homes are wired in parallel. Each socket outlet or lamp holder receives the full mains voltage and should a lamp on the lighting circuit 'blow', all the others will continue to work normally.

Worked Example 4

All the lamps in the following two circuits are identical and each is designed to work at normal brightness when the voltage across it is 4.5 V; all the cells are identical, each supplying 1.5 V.

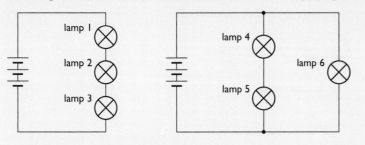

a) State the order of brightness of the lamps starting with the brightest.
b) Give reasons for your ordering in part a).
c) Explain what happens if (i) lamp 3 blows, (ii) lamp 6 blows.

a) L6, L4 and L5, L1 and L2 and L3
b) L6 is at normal working brightness as it has the necessary voltage of 4.5 V across it. L4 and L5 each have a half share, only 2.25 V. L1–L3 each have only 1.5 V each.
c) If L3 blows there is no complete path for current so all three lamps will fail to light. If L6 blows L4 and L5 will continue to light as before, unaffected by the break in the separate parallel path with L6.

Practical work

Most practical work in electricity is dealt with in the next chapter. You should be familiar with the circuit symbols in the specification and should be able to show, on circuit diagrams, the correct use of ammeters and voltmeters. Most meters used in schools and colleges will be digital, so connecting the meters the 'wrong' way round will simply result in a minus sign in front of the reading. Conventionally, electric current flows from the positive (+) terminal of the battery or power supply back to the negative (−) terminal. When using an ammeter this current should enter the + (**red**) terminal and leave by the − (**black**) terminal. When using a voltmeter the **red** lead should be connected to the point where current enters the component, the **black** lead to where current leaves the component.

[Multimeters are likely be used too – you should be able to select the correct setting for DC voltage and current measurement though you will not be expected to answer questions about specific types of multimeter.]

Chapter 10: Electrical resistance

Resistance

When a voltage is connected across a wire an electric current flows through the wire. The scientist Ohm found that, for metal wires, the current varied in proportion to the voltage provided the wire did not get hot.

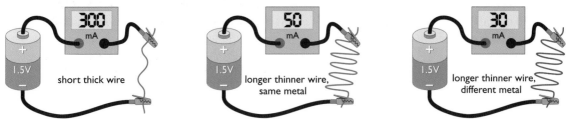

Figure 10.1

Different wires allowed different sized currents to flow for a given voltage; some wires allowed current to flow quite easily, others did not. Ohm called the property of the different wires responsible for the different currents **resistance**. This relationship is shown in the following formula:

> **voltage, V = current, I × resistance, R**

This formula can be rearranged using the triangle:

Resistance is measured in **ohms** (abbreviation, Ω). A component with a resistance of $1\,\Omega$ will pass a current of $1\,A$ when a voltage of $1\,V$ is applied across it.

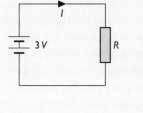

Worked Example 5

Calculate the current that will flow in this circuit if R has a resistance of **(i)** $10\,\Omega$, **(ii)** $25\,\Omega$, **(iii)** $50\,\Omega$.

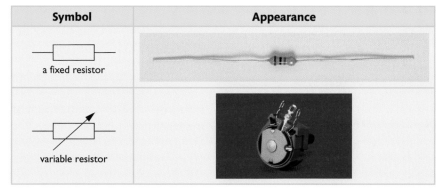

Rearrange the formula above to get $I = \dfrac{3\,V}{R}$.

So, **(i)** $I = \dfrac{3\,V}{10\,\Omega}$ therefore, $I = 0.3\,A$

(ii) $I = \dfrac{3\,V}{25\,\Omega}$ therefore, $I = 0.12\,A$

(iii) $I = \dfrac{3\,V}{50\,\Omega}$ therefore, $I = 0.06\,A$

Notice that, for a given voltage in the circuit, **the greater the resistance, the smaller the current**.

Components with resistance

Resistors

Symbol	Appearance
a fixed resistor	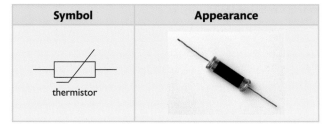
variable resistor	

Resistance can be varied by rotary or sliding action. Resistors have many uses, for example, in simple lamp dimmer circuits, for speed control in slot car games and for volume control in audio equipment.

Thermistors

Symbol	Appearance
thermistor	

Thermistors are made from semiconductors and have resistance that changes significantly with change in temperature. **Their resistance decreases as temperature increases.** (But it is possible to make thermistors whose resistance increases with temperature.) They are used in temperature sensing circuits.

Light-dependent resistors

Symbol	Appearance
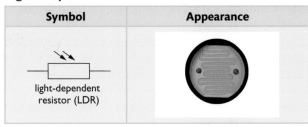 light-dependent resistor (LDR)	

Also made from semiconductor, LDRs have resistance that changes with the amount of light shining on them. **More light means less resistance**. LDRs are used in light sensing circuits, like those used to turn on street lights when it gets dark.

Diodes and light-emitting diodes (LEDs)

Symbol	Appearance
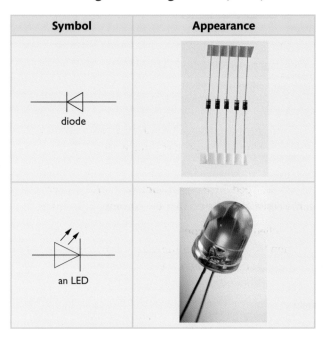 diode	
an LED	

Both diodes and LEDs only allow an electric current to pass in the direction of the arrow. In one direction they have a huge resistance to current and in the other, very little resistance. You can think of diodes and LEDs as one way streets for electric current. LEDs light up when a current passes through them and are used as indicators to show that a device is turned on. Diodes control the direction in which current can flow in a circuit and are used in power supplies for DC equipment.

Worked Example 6

State what happens in each of the following circuits when the switch is closed.

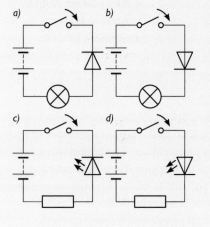

a)

b)

c)

d)

a) The lamp does **not** light.

b) The lamp **does** light.

c) The LED does **not** light.

d) The LED **does** light.

Note: LEDs are destroyed if too much current passes through them. The resistor in series with the LED is needed to limit the current.

Current-voltage (I–V) graphs

You need to know how to investigate how the current through a component changes as the voltag
connected across it is changed. This is described under practical work below. The results of such ex
are usually shown in graphs. You should be familiar with the graphs shown below.

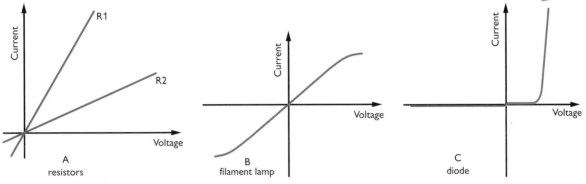

Figure 10.2

A: Resistors and wires obey Ohm's Law. Current, **I**, is proportional to voltage, **V**, and the graphs are *straight lines which pass through the origin* (0,0) of the scales. R1, the *steeper* line, is a resistor with *smaller resistance* than R2.

B: The filament in a lamp is a metal wire but it gets very hot indeed. **The resistance of a metal increases with temperature** – the graph curves when the lamp reaches its working current and temperature.

C: Diodes have very large resistance when the voltage is applied in the 'wrong' direction – this is shown by the horizontal line when the voltage is negative. When the voltage is in the 'right' direction (forward biased), when it reaches around 0.7 V the resistance drops to a small value – the graph curves and become very steep.

Practical work

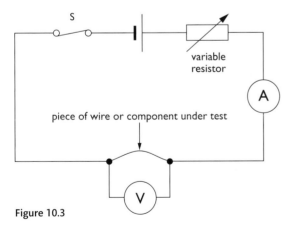

Figure 10.3

This circuit allows the relationship between current and voltage to be investigated. Changing the value of the variable resistor changes both the current through the component under test and the voltage across it. A voltmeter capable of reading up to 10–20 V is suitable but you may need to use a milliammeter if the component under test has a large resistance.

When testing a lamp the supply voltage must not be greater than the rated voltage of the lamp.

When testing resistors (demonstrating the Ohm's Law straight line relationship) the resistor (or wire) must not be allowed to get hot – this means the maximum current flowing must be kept at a suitably low value.

When testing diodes the maximum forward voltage must not be greater than 1.0 V. Most rectifier diodes can be reverse biased (cathode voltage higher than anode) to –50 V safely.

The maximum forward voltage for standard LEDs should not exceed 3.0 V (they don't start to conduct and therefore light until V_F is ~ 2 V). Reverse voltage for LEDs **must not** be more than 5 V.

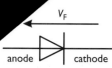

anode $\triangleright$ cathode

Figure 10.4 *A diode is* **forward biased** *when the anode is more positive than the cathode – if $V_F > 0.7\,V$ then the diode conducts*

The circuit shown on page 29 can also be used to measure the resistance of a component; measure I and V then use the rearranged Ohm's Law formula,

$$R = \frac{V}{I}.$$

Resistance meters pass a known current through a resistor then measure the resulting voltage – they are calibrated to read directly in ohms (kΩ or MΩ, as required).

1 Look at the diagrams showing the wiring of a three pin mains plug:

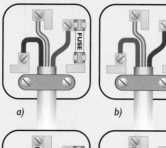

a)   b)

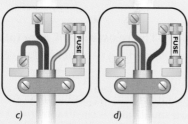

c)   d)

Which shows the **correct** wiring?

2 Look at the diagram of a plug shown below. It has a number of dangerous faults. List them. You should be able to find five.

3 What is the purpose of a fuse in a mains plug?

4 What is the purpose of the earth wire in a plug?

5 Give an example of a household appliance that is commonly double insulated. Why is an earth connection not necessary in this type of appliance?

6 List five examples of appliances that use electricity to generate heat as a useful form of energy output.

7 Some parts of an electrical circuit are designed to have very little electrical resistance but others must have resistance, for example, the flex of a reading lamp has a resistance of less than an ohm but the filament in the lamp can have a resistance of 1000 ohms. Explain, in terms of energy transfer, why this is so.

8 a) What is the purpose of a fuse in an electric circuit?

b) Fuses for electric plugs are available with the following ratings: 1 A, 3 A, 5 A, 10 A and 13 A. Do a calculation to choose the correct value of fuse to use in a plug attached to each of the following 230 V mains appliances:

   (i) a 900 W toaster,

   (ii) a 350 W electric drill,

   (iii) a 1.2 kW coffee machine.

9 a) Write down the word equation used to calculate how much electrical energy is transferred when an appliance is used for a known amount of time running from a given supply voltage if you are also told the current it draws.

b) How much energy is transferred by

   (i) a reading lamp running from a 230 V mains supply for 2½ hours drawing a current of 220 mA,

   (ii) a 12 V, 36 W car headlamp during a 50 minute journey at night?

10 A teacher says that the AC mains supply is like playing a mouth organ because you suck as well as blow to play notes. What does this tell you about an AC supply and how does it differ from a DC supply?

11 In an experiment to demonstrate static electricity, a comb which has been rubbed with a dry cloth is brought close to a stream of water trickling from a tap.

a) Explain how the comb has become charged.

Section B: Electricity

b) Pupils then try the same experiment with their rulers; most are successful but a few pupils, who have steel rulers, are not. Explain why they are unsuccessful.

c) One pupil who has watched the experiment carefully says 'Just like in inkjet printers, Miss!'. Explain this observation.

12 In another experiment the teacher has suspended two different types of plastic from nylon threads so that they can turn freely. She tells the class that the polythene strip has **negative** charge and the cellulose acetate strip has **positive** charge. She asks the class to try charging different strips of material with a dry cloth and then to bring their charged strips close to the two suspended rods. Explain the following observations made by members of her class:

a) Katie discovers that her strip attracts the polythene rod and repels the acetate rod.

b) Shazia gets the opposite result from Katie.

c) Joe finds that his strip attracts both the suspended rods.

d) Amar finds that the *cloth* he used to charge his strip of material attracts the polythene rod.

13 Copy and complete the following sentences about electric current and charge:

An electric _____ is the rate of flow of _____ in a circuit. The amp is the unit of _____ and the unit of electric charge is the _____. The relationship between current, charge and time is given by the equation _____ = _____ × _____.

14 Conventionally we say that an electric cell circulates positive charge from the positive terminal around the circuit and back to the negative terminal, as shown in the figure below.

a) What charged particles are free to move in metals?

b) What is the sign of the charge on these mobile particles?

c) In what direction do they move around the circuit?

[Note that we use conventional current in rules like Fleming's Left Hand Rule.]

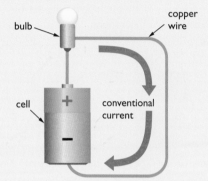

15 *a)* Copy and complete the following sentences about the volt and voltage:

Voltage is the measure of how much _____ in _____ is transferred per _____ of _____ that passes through a component.

b) The voltage across a bulb in a circuit is 12 V; how much energy is transferred to the bulb filament by each unit of charge that is passed through the bulb? What are the two main energy conversions that take place in the bulb?

16 *a)* Draw circuit diagrams to show how three lamps are connected **(i)** in **series** and **(ii)** in **parallel**.

b) One lamp 'blows' (becomes open circuit) in each type of circuit; state the effect that this has on the other lamps in each circuit.

c) Two switches are used in a blender to control when the cutting blades turn. One is closed when the lid of the blender is securely in place and the other is pressed by the user. Should these switches be connected in *series* or in *parallel*? Give a reason for your answer.

17 Here are some circuits made with lamps, resistors and cells. The cells all have the same voltage and all the lamps and all the resistors are identical to each other.

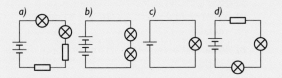

Place the circuits in order of the size of current flowing in each, from highest to lowest.

18 Here are some graphs showing how the voltage and current are related in resistors, lamps and diodes. Identify which component each graph refers to. You may need to use a component name more than once.

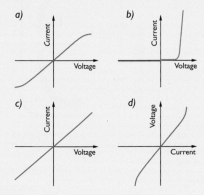

19 In each of the following circuits the resistance is changed.

a) Name the components X, Y and Z.

b) State and explain what happens to the current in each circuit as a result of the change. You can assume that the voltage of the battery is constant in each circuit.

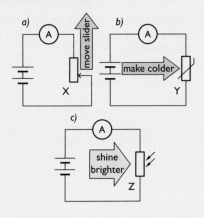

20 Write the equation that shows the relationship between *V*, the voltage across a conductor, and *I*, the current which flows through it as a result and the resistance of the conductor.

a) With *V* as the subject of the equation,

b) with *I* as the subject of the equation, and

c) with *R* as the subject of the equation.

21 Calculate the missing voltmeter reading, ammeter reading and resistor value in the following three circuits.

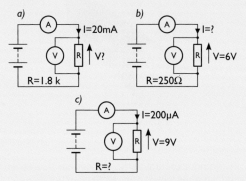

22 Here is a circuit in which a resistor is being used to protect an LED. The LED is being used as an indicator to show that the switch is is on and the circuit is powered. The LED is designed to operate with a voltage of 1.8 V across it and passing a current no greater than 12 mA.

a) What would happen if the switch was closed without *R*?

b) What is the voltage across *R* when the switch is closed?

c) Calculate the value of *R* needed to limit the current through the LED to 12 mA.

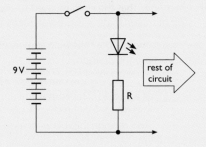

Chapter 11: Properties of waves

What are waves?

> Waves can transfer energy and information from one place to another.

Waves can be divided into two types: **mechanical waves** (waves that need a material medium to travel through: gas, liquid or solid) and **electromagnetic or EM waves** (waves that can travel through a vacuum – they do not need matter like mechanical waves).

> Energy and information are transferred by waves without transfer of matter.

Mechanical waves can be of two types: **transverse**, like ripples on a pond, and **longitudinal**, like sound waves.

Both types of waves can be demonstrated on a slinky spring.

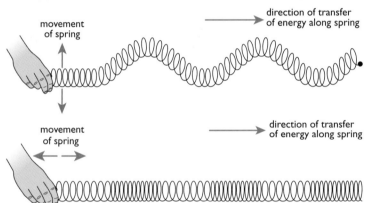

Transverse waves: In transverse waves the medium (in this case the coils) moves at **right angles** to the direction of the motion of the wave.

Longitudinal waves: In longitudinal waves the medium (in this case the coils) moves **in the same direction** as the motion of the wave.

Figure 11.1

In both types of mechanical wave, the particles that make up the medium oscillate (move back and forth) about fixed points within the medium. The difference, as shown with the slinky above, is how the direction of their movement compares with the direction of the resultant wave travelling through the medium.

Describing waves

The following diagram shows what a wave, like a ripple on the surface of water, looks like at an instant in time:

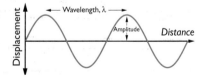

Figure 11.2 *Displacement is how far the water surface has moved from its rest position. It can move above (positive displacement) or below its rest position as in this example of a transverse wave, or in the direction of the wave movement (positive displacement) or away (negative displacement) from the direction the wave is moving in the case of a longitudinal wave. Distance is measured from the wave source.*

> **Wavelength** is the distance between corresponding points in the wave – one crest to the next crest in a transverse wave. (A more strict definition would be the distance between two adjacent points of similar displacement.)

> **Amplitude** is the maximum displacement of a part of the medium from its rest position.

Figure 11.3 shows how a particular point in the medium moves with time. **It is important not to m[ix]** these two graphs up, **so** *look at the labels* **on the horizontal axes**.

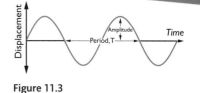

| The **period** of a wave is the time for one complete cycle of the waveform. |

| The **frequency** of a wave is the number of cycles of the waveform per second. |

Figure 11.3

The relationship between the frequency and period of a wave is:

$$\text{frequency} = \frac{1}{\text{period}}$$

Frequency is measured in hertz (Hz) provided that the period is in seconds.

Worked Example 1

1. The graph to the right shows how the displacement of a wave varies with time. Use it to find the **period** of the wave and hence calculate the **frequency** of the wave.

    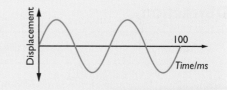

    Two complete cycles take 100 ms, so the period, T, for one is <u>50 ms</u>.

    $f = \dfrac{1}{T}$ where frequency, f, is in Hertz if T is in seconds.

    50 ms = 0.05 s

    So, $f = \dfrac{1}{0.05}$

    Therefore, $f = 20$ Hz

2. Calculate the **period** of a waveform with a **frequency** of 50 Hz.

    $f = \dfrac{1}{T}$ can be rearranged to make T the subject: $T = \dfrac{1}{f}$ where T is in seconds if f is in Hertz.

    So, $T = \dfrac{1}{50}$

    Therefore, $T = 0.02$ s

The wave equation

You need to know this equation and be able to rearrange it as required.

| **wave speed, v = frequency, f × wavelength, λ** |

Units: v in m/s, f in Hz and λ in m

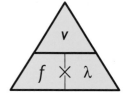

Worked Example 2

All electromagnetic waves travel with the same speed in a vacuum. This speed is 300 000 000 m/s. Radio waves are a part of the electromagnetic spectrum so they also travel at this speed. Radio station A transmits radio waves with wavelength 200 m and radio station B transmits radio waves with wavelength 300 m. Use the wave equation to calculate the frequencies of the radio waves transmitted by these two stations.

First, rearrange the wave equation to give $f = \dfrac{v}{\lambda}$

So,

For station A: $f = \dfrac{300\,000\,000}{200}$

Worked Example 2 *(continued)*

Therefore, $f = 1\,500\,000$ Hz

For station B: $f = 300\,000\,000/300$

Therefore, $f = 1\,000\,000$ Hz

Note that, for waves that travel at the same speed, the higher the frequency of a wave the shorter is its wavelength.

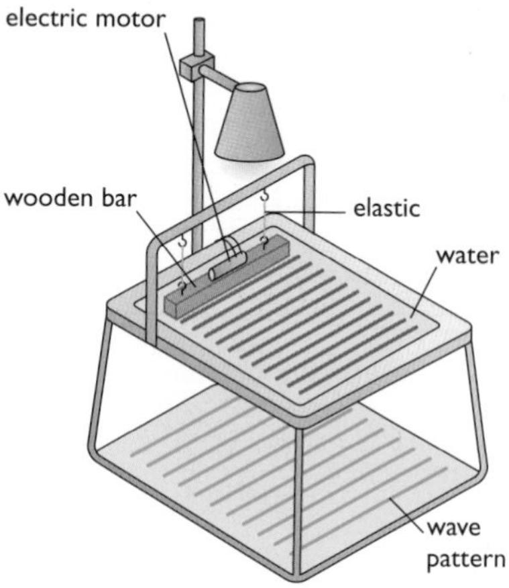

Figure 11.4 *The light shines through the water and we can see the patterns of the waves*

The ripple tank

The ripple tank is a simple piece of apparatus that makes it possible to show how waves behave. You need to be able to demonstrate some basic wave behaviour using diagrams obtained from ripple tank experiments. More details are given in the practical section.

Diffraction

Diffraction is the spreading of waves as they pass by edges of obstacles.

This is shown clearly in ripple tank demonstrations:

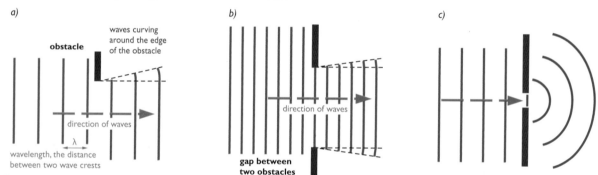

Figure 11.5

Diffraction occurs with waves of all types. The idea of **sound waves** bending round corners is unsurprising, you can hear someone calling you when you are around a corner. The suggestion that **light waves** curve around obstacles appears to contradict the obvious statement that 'light travels in straight lines'. When waves pass the edge of an obstacle they curve around the edge as shown in Figure 11.5a. The effect is small when the gap size is large compared to the wavelength as in Figure 11.5b. The effect is much greater when the gap size is similar to the wavelength as in Figure 11.5c. Sound waves have wavelengths of size about 1 cm (very high pitched sound) to 15 m (very low notes). Light wavelengths are all *smaller than one-thousandth of a millimetre*, so diffraction effects are not usually noticeable.

Practical work

You should be able to demonstrate different types of wave in a practical situation. The slinky spring can be used to show both transverse and longitudinal waves. You can also show transverse waves by shaking the end of a rope up and down, or by creating ripples on the surface of water. EM waves are transverse, but this cannot be demonstrated simply. Sound waves are a good example of longitudinal waves but, once again, you cannot actually see particles moving backwards and forwards along the line of travel as you can with the slinky.

The ripple tank makes it possible to demonstrate many properties common to all types of wave. Refraction, the bending of light waves as they pass from one material to another, can be demonstrated by reducing the

depth of water in the ripple tank (with a transparent glass or plastic sheet). Ripples travel more slowly i[n]
the depth of water in the ripple tank is smaller. When setting up a ripple tank it is therefore important t[hat]
the tank is level. Another problem with ripple tanks is unwanted reflections from the sides of the tank; t[hey]
result in pretty patterns but make analysis of what you see very difficult. Most ripple tanks have sloping
sides (beaches) to reduce unwanted reflections.

To show the interesting effects of diffraction (and interference) you need to set up continuous plane wavefronts and (circular wavefronts respectively). This is done with a vibrating bar placed either directly in contact with the water (for plane wavefronts, Figure 11.6) or with two dippers just touching the water (for circular wavefronts, Figure 11.7). The frequency of vibration and thus the frequency of the waves is controlled by varying the speed of the electric motor attached to the beam.

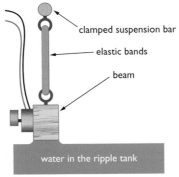

side view: Showing beam with motor to make it vibrate up and down to make plane wavefronts

Figure 11.6

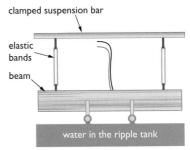

front view: Showing beam with two dippers fitted to create two sets of circular wavefronts. The beam has been raised clear of the surface.

Figure 11.7

Chapter 12: Using waves

Electromagnetic waves

Electromagnetic waves are transverse changes in the electric and magnetic fields in space; they do not use a material medium and all travel at the same speed through a vacuum (free space) – 300 000 000 m/s. They all have wave properties in common but are produced in different ways and have special properties that depend on their frequency and, therefore, their wavelength.

You should know the order of the different types of EM waves that make up the continuous **electromagnetic spectrum**:

> Radio waves – microwaves – infrared – visible light – ultraviolet – X-rays – gamma (γ) rays
> *Increasing frequency →*   *← Increasing wavelength*

A mnemonic can help: **R**un **M**iles **I**n **V**ery **U**npleasant e**X**treme **G**ames (more memorable if it is your own).

You should also be able to recall the order of the colours in the visible part of the EM spectrum:

> **RED** - ORANGE - YELLOW - **GREEN** - BLUE - **INDIGO** - **VIOLET**
> *Increasing frequency →*   *← Increasing wavelength*

The standard rhyme is: **R**ichard **O**f **Y**ork **G**ave **B**attle **I**n **V**ain (there are many others).

Since the speed, *c*, of all EM waves is **constant** you can see that a higher frequency **must** mean a shorter wavelength

It is worth emphasising that the EM spectrum is *continuous* – it is only broken up into distinct zones for convenience. For example, the visible light spectrum is made up of an indeterminate number of colours that blend smoothly from one shade to the next, as you can see if you look at the palette of colours in any word processing package.

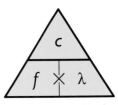

Uses and dangers of electromagnetic radiations

Figure 12.1

Uses: Radio waves are used for communication of information. This can be speech, radio and television, music and encoded messages like computer data, navigation signals and telephone conversations. Radio waves have the lowest frequencies and longest wavelengths, television and digital communication uses higher frequencies and shorter wavelengths.

Figure 12.2

Uses: Microwaves are used for heating food (by directly transferring energy to the water content in food) and also for satellite communication and radar (used to detect to aircraft and shipping).

Dangers: Microwaves can directly heat internal body tissue, so serious damage can occur before pain is felt. Microwave cookers have a cut-out which turns off the microwave radiation when the cooker door is opened.

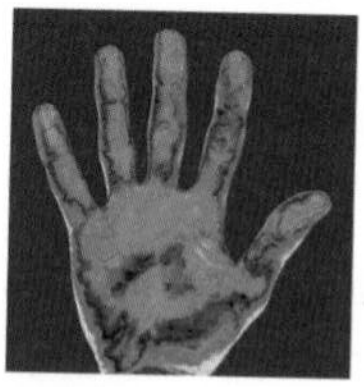

Figure 12.3

Uses: Infrared is used in heating devices and in night vision cameras, which pick up the different temperatures of objects because the wavelength of infrared from warm objects (like people and animals) is shorter than the infrared radiation from cool objects (like stones and buildings).

Dangers: Infrared is readily absorbed by our skin and will result in burns. Workers whose occupations expose them to infrared radiation wear protective clothing with reflective surfaces and good insulating properties.

Uses: Visible light is used by humans to see things, in photography and in some types of fibre optics to look inside the human body when diagnosing and treating medical conditions.

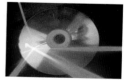

Figure 12.4

Figure 12.5

Uses: Ultraviolet is used in fluorescent lamps, disco 'black' lights and sterilising water. Some insects can see into the ultraviolet part of the spectrum and use this to navigate and to identify food sources.

Dangers: Ultraviolet is responsible for skin cancers and can damage the eyes. The sun is a main source of ultraviolet radiation and precautions against potentially dangerous exposure include the use of high protection factor suntan cream and covering up. Sunglasses will reduce the exposure of the eyes to UV.

Uses: X-rays are used to examine the internal structures of the body in medical diagnosis and to investigate the crystal structure of materials used in manufacturing. The X-rays produced by collapsing stars are also used in radio astronomy.

Dangers: X-rays can produce cell mutation and cancer. Women are advised to avoid having X-rays during pregnancy. Hospital worker and others who routinely use X-rays wear protective clothing and operate X-ray machines remotely to reduce exposure.

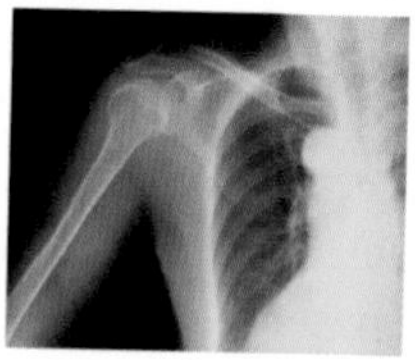

Figure 12.6

Uses: Gamma rays are used to treat certain types of cancer. They are also used for sterilising medical equipment and food products.

Dangers: Gamma rays also cause cell mutation and cancers. Workers in the nuclear industry wear badges to monitor long term exposure and either wear dense protective clothing or use remote handling devices when dealing with radiation sources.

Figure 12.7

Analogue and digital signals

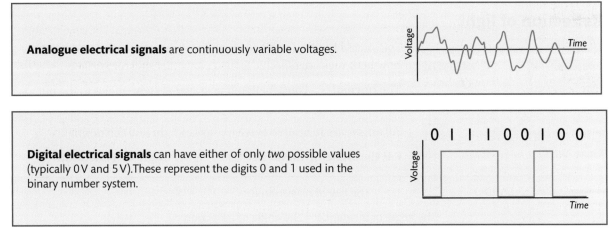

Analogue electrical signals are continuously variable voltages.

Digital electrical signals can have either of only *two* possible values (typically 0 V and 5 V).These represent the digits 0 and 1 used in the binary number system.

Advantages of digital transmission

Signals transmitted over large distances have two main problems: they lose energy so the signals get weaker (jargon: the signals are *attenuated*) and they pick up unwanted *interference* or **noise**, that is unwanted signal.

To overcome the problem of the analogue signal getting weaker as it travels it must be **amplified** to boost the energy of the signal at points along its path. (Amplification is the opposite of attenuation.) Unfortunately both the wanted signal and the unwanted noise are amplified, so the quality of analogue signals is reduced.

Digital signals can be regenerated, both amplifying *and* restoring their distinct '0' and '1' shape electronically thus virtually eliminating the unwanted noise from the signal.

Optical fibres allow a much wider bandwidth. This means that many different digital signals can share the same optical fibre, so much more information can be transmitted along an optical fibre than by using an analogue signal.

Practical work

There is not much practical work for this chapter. You should be familiar with splitting white light up into the visible spectrum (rainbow) using a triangular glass prism. To obtain a good spectrum you need a darkened room, a white screen on which to project the spectrum, a narrow slit of light produced from a white light source (a ray box with a hot filament lamp) and a good

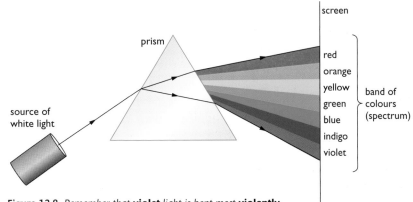

Figure 12.8 *Remember that* **violet** *light is bent most* **violently**

prism. Varying the angle at which the light falls onto the prism will allow you to get the best spreading effect. Refraction is dealt with in the next chapter.

Chapter 13: Light waves

Reflection of light

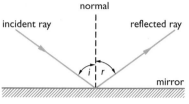

Figure 13.1

You should be able to able to draw this diagram which demonstrates how light rays are reflected from a plane (flat) mirror surface.

The **normal** is a construction line drawn at right angles to the mirror surface where rays of light strike it.

All angles are measured between rays of light and the normal.

i is the **angle of incidence** and r is the **angle of reflection**

The law of reflection is:

the angle of incidence = the angle of reflection

Forming an image in a mirror

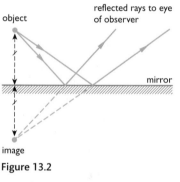

Figure 13.2

You should be able to draw a ray diagram to show how an image is formed in a plane mirror.

The image formed in a plane mirror is a **virtual image** as it appears to be behind the mirror; rays of light are not *actually* coming from the place where the image seems to be. *The image appears to be the same distance behind the mirror as the object is in front of it.*

When drawing this diagram real rays of light are solid, but the construction lines behind the mirror are dotted, showing where the rays of light that are reflected *appear* to have come from.

You should make sure that your diagram shows rays that reflect according to the rule above ($i = r$)!

Refraction of light

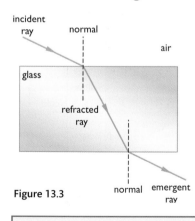

Figure 13.3

Refraction of light is the bending (change in direction) that happens when light travels from one material, like air, into another, like glass.

Light travels more slowly in substances like glass, perspex and water.

*When a light ray travels from air **into** glass (or similar substances) it is bent **towards** the normal, and when a light ray travels **out of** glass back into air it is bent **away** from the normal.*

The degree of bending depends on the material that the light is travelling into (and out of). The property of the material that determines the amount of refraction is called the **refractive index, n**, of the material. The bigger the value of n the greater the bending effect.

The law of refraction is:

$$\text{refractive index, } n = \frac{\sin i}{\sin r}$$

The angle of incidence, *i*, is measured between the ray of light in air and the normal. The angle of refraction, *r*, is measured between the ray of light in the material and the normal (see Figure 13.4). An experiment to measure *n* is described in the practical section on page 43.

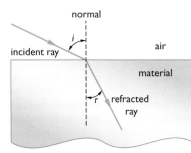

Figure 13.4

Worked Example 3

In an experiment to measure the refractive index of glass, Mariam measures the angle of incidence and angle of refraction for a given ray of light passing from air into a glass block. Her results are: *i* = 50° and *r* = 31°. Use her results to calculate the refractive index of the glass to two significant figures.

Use: $n = \dfrac{\sin i}{\sin r}$

So, $n = \dfrac{\sin 50}{\sin 31}$

Therefore, n = 1.5

Note that refractive index has no units.

Total internal reflection

The following diagrams show an experiment to demonstrate total internal reflection (TIR):

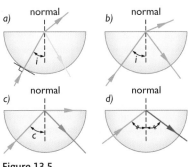

Figure 13.5

Figure 13.5a shows a ray of light directed toward the midpoint of the diameter a semicircular block of glass. The ray passes across the curved face of the block without changing direction because it meets the surface at 90° When the ray meets the straight surface it is refracted away from the normal because it is travelling from glass to air. You will also notice a small amount of reflection occurs, shown by the faint red line.

Figure 13.5b shows the same, but here the size of angle *i* is greater. The ray emerges from the flat surface having been refracted through a greater angle and you will notice that the proportion of the ray reflected has increased.

Figure 13.5c shows the ray of light meeting the flat surface at the **critical angle**, **c**. Now the ray just emerges from the block at an angle of 90° to the normal. The amount of light reflected is increased again. If you do this experiment with a ray of white light the ray will show dispersion into the colours of the spectrum.

Figure 13.5d shows the ray meeting the flat surface at an angle *greater than the critical angle and it is totally internally reflected*; <u>no</u> light emerges from the flat surface of the block.

Refractive index, $n = \dfrac{1}{\sin c}$

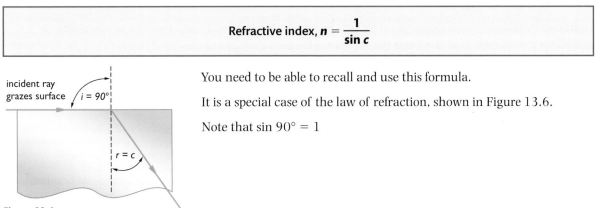

Figure 13.6

You need to be able to recall and use this formula.

It is a special case of the law of refraction, shown in Figure 13.6.

Note that sin 90° = 1

Uses of total internal reflection

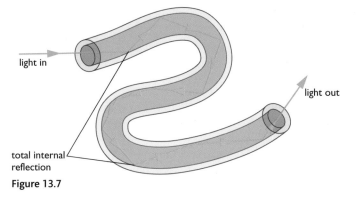

light in

total internal reflection

light out

Figure 13.7

You should be able to explain the use of TIR in **optical fibres**. Remember that *total internal reflection* *only* occurs when light is travelling from a material in which it travels more slowly, towards a medium in which it travels more quickly (e.g. travelling through glass towards a boundary with air). This is why the outer layer of an optical fibre has a *lower* refractive index than its core. Optical fibres are now widely used for digital communication between computers and for speech and video communication. They are also used in medicine for 'keyhole' surgery in devices called **endoscopes**.

Worked Example 4

1 Show that the critical angle for glass with a refractive index of 1.5 is 42° to two significant figures.

Use: $n = \dfrac{1}{\sin c}$

Rearrange this to give: $\sin c = \dfrac{1}{n}$

So, $c = \sin^{-1}\left(\dfrac{1}{1.5}\right) = 42$

Therefore the critical angle is 42°

2 Use this to explain how glass prisms are used in a periscope. Your answer should include a labelled diagram.

The periscope uses two right angled prisms with two 45° angles as shown. Rays of light from the object being viewed enter the top prism at (or close to) 90° to the face and suffer no (negligible) change in path. This means that the rays meet the long face of the prism with an angle of incidence of 45° – this is greater than the critical angle of 42° so total internal reflection occurs and the ray path is turned through 90°. This is repeated by the second prism and rays then pass to the viewer's eye. As *two* reflections have occurred the final virtual image is not *laterally inverted*.

Note: *laterally inverted* is jargon for back to front, that is, left–right reversed.

two 45° isosceles triangles

Diffraction of light

Light **can** be diffracted, but the physical set up to demonstrate this is difficult. For it to be observable, a special light source must be used and the slit through which the light passes must be very narrow. We have seen above that the extent of the spread of a wave is small if the size of the gap is large, and the wavelength of orange light (for example) is roughly 0.6 millionths of a metre. A slit just 0.6 mm wide is 1000 times bigger that the wavelength.

Practical work

You should be able to describe the experiment shown in the four figures of the block of glass on page 41 and be able to measure the critical angle, **c**. In this experiment shine a ray of light so that it does not bend on entering the curved face and it just leaves the block as shown in Figure 13.8. Draw round the block and mark the points that the rays enter and leave the block as carefully a possible. *Remove* the block and join up points A and B with a pencil. Use a protractor to draw the normal at B and then measure the critical angle. It is difficult to measure c to better than ± 2° particularly with white light, which will disperse into the colours in the spectrum of visible light. Use of a green filter may improve your ability to locate the point where the ray is *just* totally internally reflected.

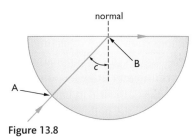

Figure 13.8

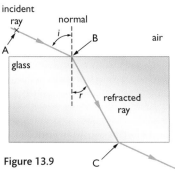

Figure 13.9

You should also be able to measure the refractive index of glass given a rectangular block as shown in Figure 13.9. As above shine a ray of light through the block, draw round it and mark the points B and C where the ray of light enters and leaves the block. Also carefully mark a point A on the incident ray. Remove the block then join A to B and B to C with straight pencil lines. Use a protractor to mark in the normal at B and measure the angle of i.

Angle of incidence is **i**, and the angle of refraction, **r**.

Now use the formula **n = sin i / sin r** to calculate **n**.

You should repeat this experiment for a range of values of **i** from 30° to 70° and find an average value for **n**.

Chapter 14: Sound

Sound waves

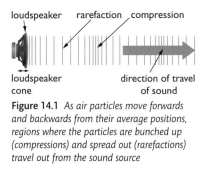

Figure 14.1 *As air particles move forwards and backwards from their average positions, regions where the particles are bunched up (compressions) and spread out (rarefactions) travel out from the sound source*

Sound travels as a **longitudinal wave** through gases, liquids and solids. It is useful to know that the speed of sound in gases is generally lower than in liquids and slower in liquids than in solids. You do not need to recall the following figures, but they show this point: in air sound waves travel at 340 m/s; in water at 1500 m/s; in steel at 5000 m/s. [These are all approximate and will vary with temperature and other variables.]

Sound waves *cannot* travel through a vacuum.

The range of frequencies that the human ear can detect is from **20 Hz to 20 kHz**; these frequencies are referred to as **audio frequencies**.

Worked Example 5

Sound waves travel at 340 m/s in air. Calculate the maximum and minimum wavelength of audio frequency sounds in air.

Rearrange the wave equation to give $\lambda = \dfrac{v}{f}$

Longest wavelength: $\lambda = \dfrac{340}{20}$

So, $\lambda = 17$ m

Shortest wavelength: $\lambda = \dfrac{340}{20\,000}$

So, $\lambda = 0.017$ m

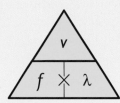

Note: For waves that travel at the same speed, the higher frequency waves have shorter wavelengths.

Sound waves share the properties of reflection, refraction and diffraction common to all types of waves.

Echoes show waves reflecting. Refraction is harder to demonstrate, but a feature of refraction, total internal reflection, is easily demonstrated with sound waves. Remember that TIR occurs when waves are travelling in a medium and strike a boundary with a medium in which waves travel faster. The whispering gallery at St Paul's Cathedral in London is an example of total internal reflection of sound. Stethoscopes also use this property – the hollow tube is the sound wave equivalent of an optical fibre. Diffraction occurs readily through gaps like doorways because a typical wavelength of sound ~ 0.3m is of a similar order of magnitude to the width of an average door space.

Measurement of the speed of sound waves

See the practical section below.

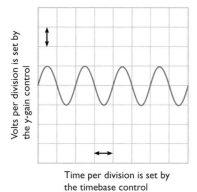

Time per division is set by the timebase control

Figure 14.2

Pitch and loudness

The pitch of any sound depends on how quickly the sound producing system (e.g. vocal chords, string on a violin) is vibrating. The vibrating system causes sound waves with the same frequency. A microphone converts sound waves into electrical signals; small voltages that vary with time in the same way as the pressure of the air changes around the microphone as sounds reach it. An oscilloscope amplifies these voltages (by a factor controlled by the y-gain on the oscilloscope) and produces a graph which shows how the voltages (and hence the pressure variations in the sound waves) are changing with time. A control on the oscilloscope called the timebase allows you to set and change the timescale on the graph.

> The height of the oscilloscope trace (its amplitude) depends on the **loudness** of the sound. The louder the sound the greater the amplitude.
>
> The **pitch** of a note depends on the frequency of the sound and since $T = 1/f$ the time for one complete cycle of the wave gets shorter as the note becomes higher in pitch – *the more waves on the screen the higher the note.*

Worked Example 6

The traces shown in the figure were all obtained with the same microphone – the same oscilloscope was used and the oscilloscope settings were not altered.

Put the three results in order of:

a) Increasing loudness

b) Increasing pitch.

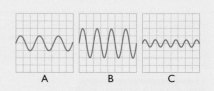

a) C   A   B

b) A   B   C

Practical work

You should be able to describe a method for measuring the speed of sound in air (v). Give details of the apparatus you would use, the measurements you would take, how you would use them to calculate speed and what techniques you would use to improve the measurement accuracy. Some experiments are described in the Student Book (pages 119–120). Here is another experiment:

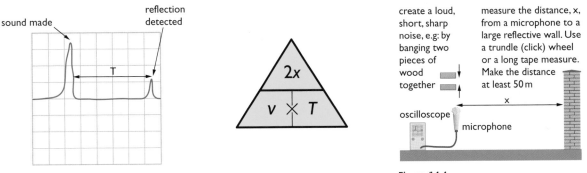

Figure 14.3

Figure 14.4

The oscilloscope time per division should be set to a value appropriate to the expected time for T; if $x = 50\,\text{m}$ the total distance travelled by the sound from the source and back to the microphone will be $100\,\text{m}$. The expected time will therefore be $\sim 0.3\,\text{s}$. A timebase setting of $50\,\text{ms/division}$ (so $300\,\text{ms}$ means six divisions horizontally on the screen) will be suitable. Speed, $v =$ distance travelled $(2x)$/time taken has been included with the symbols shown in this diagram.

Warm-up questions

1 Vinay makes circular ripples in a ripple tank with a dipper at A driven by a low frequency oscillator. Circular ripples spread out as shown (crests are shown as pale blue lines, troughs as dark blue lines). Three small pieces of cork are placed on the water surface at points X, Y and Z.

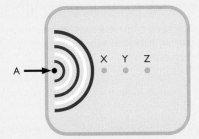

Vinay makes the following observations:
- The ripples spread out with the same speed in all directions as semicircles;
- The corks bob up and down as the waves pass through the points X, Y and Z;
- The up and down movement is noticeably smaller at Z than at X.

a) What type of wave is travelling across the tank?

b) What does the first observation tell Vinay about the way he has set up the ripple tank?

c) How does his experiment make it clear that water is not travelling across the tank, but that energy is?

d) Why is the up and down movement at Z smaller than at X?

2 Charmilie is observing traffic moving along a motorway, coned off to just one lane, from a bridge. She notices that the traffic tends to bunch up and then spread out as the traffic crawls past. This reminds her a little of a particular type of wave. Which type is it? Name another example of this type of wave.

3 a) Explain the following wave terms, using clear labelled diagrams to help your explanations:
   (i) Wavelength (ii) Frequency (iii) Period.

b) State the correct units for each of the above quantities.

c) How can the speed of a wave be calculated from *wavelength* and *frequency*?

d) How is the *frequency* of a wave related to its *period*?

4 a) What are the upper and lower frequency limits for sound waves that humans can hear?

b) Sound waves travel at about 340 m/s in air and 1500 m/s in water. Calculate the wavelengths of the highest and lowest audible sounds for sound waves travelling in (i) air and (ii) water.

5 Copy and complete the diagram overleaf showing how ripples on a ripple tank behave when they pass through a gap in a barrier.

a) What is the name of this effect?

b) How would the waves emerging on the right of the barrier differ if

(i) the wave frequency was increased,

(ii) the depth of the ripple tank was decreased,

(iii) the gap size was decreased?

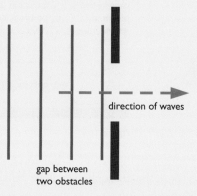

direction of waves

gap between
two obstacles

6 The diagram below shows the electromagnetic spectrum. Complete the diagram by writing the missing word in each box.

Radio waves – [ ] – infrared – [ ] – ultraviolet – X-rays – [ ]

Increasing frequency →   ← Increasing wavelength

7 a) State two features common to all the different types of electromagnetic wave.

b) State two differences between infrared waves and X-rays.

8 State the principal *seven* colours that make up the rainbow, starting with the longest wavelength.

9 *Radio waves* are used for communications, radio and television; they have no major detrimental effect. For the remaining groups of waves in the electromagnetic spectrum give one use and one health risk.

10 What kind of waves are light waves? In what ways do they differ from the waves made on a rope by moving the free end up and down?

waves on a rope made by moving
the free end up and down

11 a) The diagram below shows light reflecting from a plane mirror.

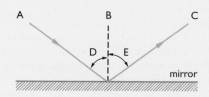

mirror

State which letter on the diagram shows:

(i) The normal

(ii) The incident ray

(iii) The reflected ray

(iv) The angle of incidence

(v) The angle of reflection

b) State the law of reflection.

12 The drawing shows a lamp on a stand in front of a mirror.

a) Copy and complete the drawing, marking two rays of light from each of the points marked with a star, to show how and where the image of the lamp is formed in the mirror.

b) What type of image is formed in a mirror?

13 a) Copy and complete the diagram to show the path of the ray of light as it travels from air into water.

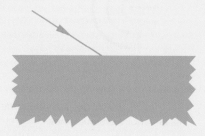

b) State the equation for the refractive index of a substance.

c) Label your diagram to identify the terms you have used in the equation.

d) Show, by drawing an extra labelled line on your diagram, the path the ray of light would have taken if it had entered a material with a larger refractive index than water.

14 Glass has a refractive index of 1.5. If a ray of light makes an angle of 75° to the normal at the point where it meets the surface of a block of this glass, calculate the angle the ray makes with the normal as it travels through the block.

15 *a)* What is *total internal reflection*?

　　 b) What is the critical angle for a perspex block with a refractive index of 1.4? Show your working.

16 *a)* Complete the four diagrams below by sketching the expected path of the ray of light as it passes through the blocks:

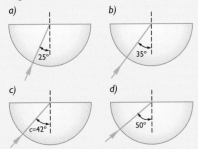

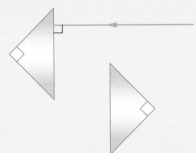

　　 b) Redraw block c) sketching the expected path of the ray of light if the refractive index of the block was **(i)** larger and **(ii)** smaller.

17 If the glass in the prisms shown has a critical angle of 40° complete the diagrams to show how the ray of light continues.

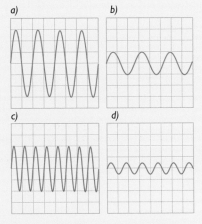

18 *a)* Explain, with the aid of a labelled sketch, how optical fibres can be used to transmit light.

　　 b) Give two practical uses of optical fibres.

19 Give examples from life that demonstrate that sound waves exhibit the following wave properties:

　　 a) reflection, and *b)* diffraction. [Note: sound waves are refracted too, but examples are less common and sometimes less obvious.]

20 Which of the following statements about sound waves is true.

　　 A: Sound waves are longitudinal waves that have frequencies from 200 Hz to 20 kHz.

B: Sound waves are transverse waves that have frequencies from 10 Hz to 10 kHz.

C: Sound waves are longitudinal waves that have frequencies from 20 Hz to 20 kHz.

D: Sound waves are transverse waves that have frequencies from 200 Hz to 20 kHz.

21 A pistol is fired 500 m away from a cliff face and an electronic millisecond timer is started by the sound. The reflected sound is detected by a microphone and this signal is used to stop the timer.

The timing is repeated several times and the average time between creating the sound and detecting the echo is 2.971 s.

　　 a) What is the speed of sound in air given by this experiment? Show your method.

　　 b) Why is it necessary to use an electronic timing system?

　　 c) Give one advantage and one disadvantage of conducting the experiment at a greater distance from the cliff.

22 Use the speed of sound you calculated in question 21.

You see a lightning flash during a thunder storm and time how long it is before you hear the consequent crash of thunder. The interval is 5.0 seconds.

How far away did the lightning flash occur?

What assumption have you made?

23 The following traces show sound waves that have been picked up by a microphone and displayed on an oscilloscope screen. (Same oscilloscope setting in each case.)

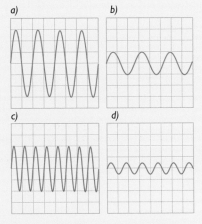

　　 a) Which of the following lists shows the sound waves in descending order of loudness?

　　　 (i) a) b) c) d)　　 **(ii)** c) b) d) a)

　　　 (iii) d) b) c) a)　　 **(iv)** a) c) b) d)

b) Which of the following lists shows the sound waves in ascending order of pitch?

(i)   a) b) c) d)   (ii)   b) a) c) d)

(iii) d) c) a) b)   (iv) b) a) d) c)

24 An oscilloscope, with time/div set to 2 ms/div and the y axis to 5 mV/div, displays the sound waveform shown here.

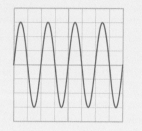

Use this information to calculate:

a) the frequency of the sound, and

b) the amplitude of the sound.

Show your method for each calculation.

Chapter 15: Energy transfers

Energy transfers

> *Key ideas*: **Energy is conserved**
>
> Energy is neither created nor destroyed, it is converted from one form to another.
>
> In any closed system the total energy (in all its forms) remains the same.

> *Types of energy you need to know of*:
>
> Thermal (heat) energy, light, electrical, sound, kinetic (movement) energy, chemical energy, nuclear energy, gravitational potential energy, elastic potential energy

When we talk of energy lost, or wasted, we mean that the energy has escaped from the closed system, for example, when heat escapes from a house. 'Wasted' energy can mean 'lost' from the system or converted into unwanted forms, like the noise produced by a motor when what we want is kinetic energy.

Worked Example 1

a) State the *main type of wanted or useful energy* output from the following devices:

    (i) a reading lamp, **(ii)** a motor car, **(iii)** a hairdrier, **(iv)** a radio, **(v)** a bicycle dynamo.

b) State a type of *unwanted or wasted energy* output from each device.

c) The hairdrier produces more than one type of useful energy output; state a second useful type of energy produced by a hairdrier.

a) **(i)** light, **(ii)** kinetic energy, **(iii)** heat, **(iv)** sound, **(v)** electricity.

b) **(i)** thermal energy, **(ii)** noise, **(iii)** noise, **(iv)** thermal energy, **(v)** thermal energy.

c) Hairdriers also produce kinetic energy – a motor drives a fan to blow air over a heating coil.

Worked Example 2

State all the energy transfers that take place when:

a) You drop a brick onto the ground; *b)* you strike a match; *c)* you ride a bike with a dynamo and turn on the headlamp.

a) The brick starts off with GPE (gravitational potential energy). As it falls, this is converted to kinetic energy (KE) and by the time it hits the ground *all* the GPE has been converted to KE. This energy is then converted to sound and thermal energy on hitting the ground.

b) You provide some KE as you drag the match head across the striking paper and this is converted to thermal energy; this heat starts the chemical reaction that converts the chemical energy into thermal energy and light (and a tiny amount of sound).

c) You convert chemical energy (from your food) into KE, some of which spins the dynamo. The dynamo converts the KE into electricity and then the electricity transfers its energy to thermal energy and light in the lamp.

There is not enough space here to give every possible device or energy transfer you may meet in an exam question. Practise as many questions of this type as you can from the Student Book, the questions at the end of this section and past papers. Although the specification is new (2009) this type of question has been examined in the old specification (2003) so there will be plenty for you to try.

Energy diagrams

Energy or **Sankey diagrams** show the energy transfers that take place in a system, like a motor car. They should show that energy is conserved; *all* the energy put into a system comes out, but in different forms – you can see this balance in Figure 15.1.

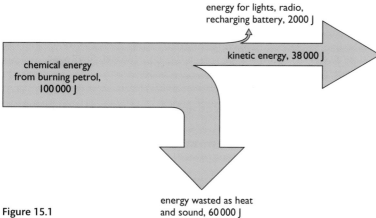

Figure 15.1

When you are asked to draw an energy transfer diagram, you will be given some figures for total energy input and/or the types of energy to which this is transferred or converted. You will then need to calculate the amount of energy needed to balance the equation: *total energy input = total energy output* in order to show that total energy is conserved. You may be asked to state the type of energy that has *not* been shown and you will probably be asked to draw (or more likely complete) a Sankey diagram. The size of the arrows should be roughly in proportion to the amount of energy they represent.

The electrical energy supplied to a lamp during a period of 5 hours is 1 800 000 J. The lamp converts 1 700 000 J to heat. What is the other type of energy produced by the lamp? How is energy converted to this form? Draw a labelled Sankey diagram to show the energy transfers that takes place.

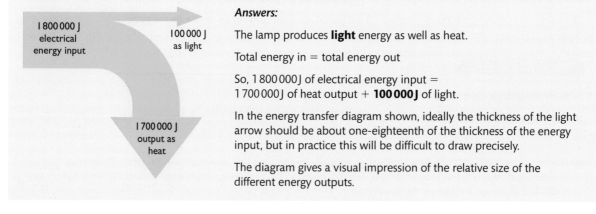

Answers:

The lamp produces **light** energy as well as heat.

Total energy in = total energy out

So, 1 800 000 J of electrical energy input = 1 700 000 J of heat output + **100 000 J** of light.

In the energy transfer diagram shown, ideally the thickness of the light arrow should be about one-eighteenth of the thickness of the energy input, but in practice this will be difficult to draw precisely.

The diagram gives a visual impression of the relative size of the different energy outputs.

Efficiency

$$\text{efficiency} = \frac{\text{useful energy output from the system}}{\text{total energy input to the system}}$$

No system can ever deliver more energy output than the amount of energy put into the system. (This would break the law of conservation of energy.)

This means that the efficiency given by the equation above *cannot be greater than 1*. In practice it will be somewhere between 0 and 1. Perfectly efficient systems (ones that do not have unwanted types of energy transfer) do not exist, and to devise a system that had no useful energy output would be pointless!

Efficiency has no units as it is a ratio of energies. Sometimes efficiency is given as a percentage:

$$\text{efficiency} = \frac{\text{useful energy output from the system}}{\text{total energy input to the system}} \times 100\%$$

So a system that is 50% efficient converts 50% of the total energy input into useful energy output.

Worked Example 4

To the nearest whole number, what is the percentage efficiency of the lamp in example 3?

Use: $\text{efficiency} = \dfrac{\text{useful energy output from the system}}{\text{total energy input to the system}} \times 100\%$

So, $\text{efficiency} = \dfrac{100\,000}{1\,800\,000} \times 100\% = 5.55\%$

Therefore, to the nearest whole number efficiency = 6%

Practical work

There is little to be done relating to this chapter, but observing processes and systems and working out what energy transfers take place and how efficient the system is at producing the wanted form of energy output is good practice for questions.

Chapter 16: Thermal energy

This chapter is about the ways in which energy is transferred from place to place in the form of *heat* and how you can reduce unwanted heat transfers. It is important to remember that heat transfer is driven by temperature difference – there will always be a net transfer of heat from places at higher temperatures to places at lower temperatures.

You will be reminded that there are three ways in which heat transfer can take place, two of which can only occur in a material medium – that is, through matter rather than a vacuum (the absence of matter).

Conduction

Thermal conduction is the transfer of thermal (heat) energy through a substance without the substance itself moving.

copper rod

wax melting

HEAT

A copper rod is coated with wax and heated at one end. As the heat energy is conducted along the rod the wax melts away as shown.

Figure 16.1 *Heat travels along the metal rod by conduction*

You will be reminded in later chapters that, in solids, liquids and gases an increase in temperature means an increase in the average amount of movement energy of the molecules of the substance. In solids, when one end is heated the vibrational energy of the molecules increases, and this energy is transferred from molecule to molecule along the material. This is a slow process in most materials, but the process is much faster in metals because some of the electrons are free to move and can carry energy through the metal much more quickly.

Convection

> **Thermal convection** is the transfer of heat energy through fluids (liquids and gases) by the upward movement of warmer, less dense, regions in the fluid.

When a gas or liquid is heated it expands. (As most gases and liquids have poor thermal conduction this effect happens closest to the source of heat.) The expansion means that the density of the fluid gets less so the warmer fluid floats upwards as it is pushed out of the way by colder, and therefore denser, fluid.

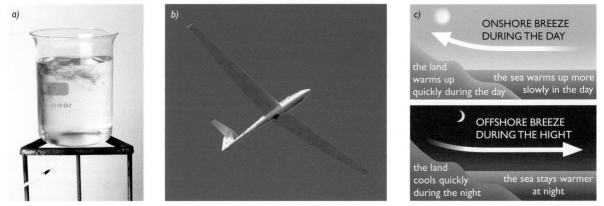

a)

b)

c) ONSHORE BREEZE DURING THE DAY

the land warms up quickly during the day

the sea warms up more slowly in the day

OFFSHORE BREEZE DURING THE HIGHT

the land cools quickly during the night

the sea stays warmer at night

Figure 16.2 *a) The purple dye shows convection currents in the water b) Gliders gain height on 'thermals' c) Onshore and offshore breezes are caused by convection*

Worked Example 5

Explain why the heating element in a kettle is placed as close to the bottom as possible.

Water is a poor thermal conductor. If the heater was at the top, only that layer of water would heat quickly, and, being less dense, would float on top of the colder, denser water at the bottom. Placing the heating element at the bottom means that as the water in contact with it warms up, the water expands and becomes less dense and is pushed upwards by colder, denser water which sinks lower from above. This results in a circulating current which heats up all of the water much more quickly.

Radiation

> **Thermal radiation** is the transfer of heat energy in the form of infrared (IR) waves. It is the *only* method of heat transfer through a vacuum.

Figure 16.3 *The Earth is warmed by IR rays from the Sun*

IR waves are part of the EM spectrum. They travel at the speed of light in space and almost as fast through air. Radiant heaters are often placed high up in a room because heat can travel in any direction – the heat from radiant heaters is felt almost instantly when the heating element starts to glow.

Energy efficient houses and insulation

Insulation aims to reduce the rate of heat transfer between areas at different temperatures. Sometimes to keep heat in (in a house, a sleeping bag or an oven) and sometimes to keep heat out (a vacuum flask, a refrigerator).

How to reduce heat transfer by conduction:

Figure 16.4

- Use a vacuum: Conduction needs matter; used in vacuum flasks, some types of double glazing, etc.

- Use air: Air is a good insulating material. Many materials like wool, feathers, fur, etc. trap air so it cannot circulate. This works because air is a very poor *conductor* of heat (like most gases). Houses use fibre glass insulation (to trap air) and cavity walls are sometimes filled with a foam (again, to stop circulation by convection).

- Use water: Wetsuits trap a layer of water around the body because water (like most liquids) is a poor conductor.

How to reduce heat transfer by convection:

Figure 16.5

- Use a vacuum: Convection needs gases or liquids; used in vacuum flasks, some types of double glazing, etc.

- Use trapped gas or liquid: This restricts circulation, which is necessary for convection to occur. The size of gap between the sheets of glass is a compromise. A narrow gap makes the effect of convection smaller, but it allows a greater amount of heat transfer by conduction.

How to reduce heat transfer by radiation:

Figure 16.6

- Use shiny surfaces: Very shiny surfaces reflect infrared (heat) radiation well. Fire fighters wear shiny suits to stop heat radiation getting to their bodies. Shiny surfaces are also poor radiators of heat. Space blankets, for example, retain the body heat of athletes or hill-walkers suffering from exposure. This is because they have a shiny inner surface which reflects heat back to the person and also a shiny outer surface which is a poor radiator of infrared.

Worked Example 6

a) Radiators in central heating systems are often painted gloss white. Is this a good idea?

b) People often hang towels, etc. over radiators to dry them off. Is this a good idea?

c) As radiators are often positioned on outside walls (not on the outside!) people sometimes stick aluminium foil behind the radiator, why?

a) The best colour for a radiator would be matt black as this radiates heat into the room most efficiently (matt black: best absorber and radiator, shiny metallic: worst absorber and radiator). Most people prefer a light gloss finish. Although, since the main way in which 'radiators' transfer heat into the room is by convection, this is not a significant problem.

b) Covering radiators prevents the circulating convection currents working efficiently, so this is not a good idea (unless your priority is to dry your towels, etc. rather than heat the room).

c) Heating the wall behind the radiator will increase the amount of heat that is conducted through the wall to the outside, which is 'lost' or wasted heat because it is not helping to keep the inside of the house warm. Covering the wall behind the radiator with a highly polished metal surface reflects heat back towards the radiator and reduces the amount of heat lost by conduction through the wall.

Practical work

There are several experiments that show the mechanisms of heat transfer and how heat losses from hot bodies can be reduced.

a)

two identical blocks of metal are heated to the same temperature. One has a shiny polished surface, the other is painted matt black.

b)

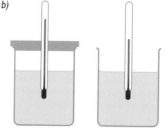

two beakers with the same mass of water initially at the same temperature. One has a lid, the other does not.

c)

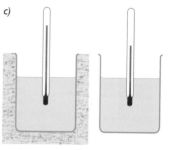

two beakers with the same mass of water initially at the same temperature. One is lagged, the other is not.

Figure 16.7

The pairs of apparatus examine just one difference in each case.

The difference in the outer surface in Figure 16.7a should show that the matt black vessel radiates heat better than the shiny one.

In Figure 16.7b the lid reduces the amount of convection above the hot liquid and therefore cools more slowly than the uncovered can.

In Figure 16.7c lagging reduces heat loss by convection.

A comparison between the three sets can show which heat loss mechanism is the most significant.

With any heat experiment it is necessary to ensure that the conditions do not vary for any of the different apparatus. For example, watch out for draughts from a window, or for surfaces with different conducting properties. Also, remember to stir when taking temperatures of liquids in order to get an even temperature distribution.

Chapter 17: Work and power

Energy and work

> **work, W = force, F × distance, d**

This is how you calculate **work** done, *W*, in joules when a force, *F*, in newtons is applied through a distance, d, in metres *in the direction of the force.*

Doing work involves transferring energy.	Energy and work are measured in joules.	Energy is the ability to do work.

What energy transfers take place in the following situations?

a) You are riding your bicycle.   **b)** You are lifting a box onto a shelf.   **c)** You apply the brakes on your bike to stop.

a) You apply a force through a distance while pedalling to work against forces such as air resistance in order to keep the bicycle moving. Your stored energy (chemical) is transferred to the movement (kinetic energy) of your legs doing work to keep the bicycle moving.

b) You are applying a force to lift the box through a distance. The work you do transfers energy to the box in the form of increased gravitational potential energy (GPE).

c) When you brake, work is done by the force of friction between the brake pads and the wheel. The kinetic energy is transferred largely to heat as the brakes and wheel rims heat up.

Gravitational potential energy

GPE = mass, *m* × gravitational field strength, *g* × height, *h*

Reminder: Gravitational field strength is the force per kilogram acting on an object in a gravitational field, units N/kg (g is numerically equal to the acceleration due to gravity).

In order to lift an object you must apply enough force to overcome gravity. The force of gravity on an object is called its weight and this is defined as:

weight = mass × gravitational field strength **or** weight = *mg*.

Since work = force × distance and distance in this case is the height, *h*, you can see how **GPE = *mgh***.

The units of GPE will be in *joules*, provided the mass is in *kg*, the gravitational field strength is in *N/kg* and the height is in *metres*.

The energy is stored as gravitational potential energy. This energy can be *returned* by releasing the object from its new position, thereby allowing gravity to *do work* on the object and causing it to speed up and gain kinetic energy (KE).

Figure 17.1 shows a roller coaster ride. At the beginning of the ride motors drag the cars to the top of the highest part of the ride. As this happens electrical energy → movement energy → GPE.

As the car swoops down the track the GPE is converted to KE or movement energy. It would continue moving around the ride without stopping until it reached the highest point once again without needing any more energy input – if energy was not converted to other forms along the way.

As the car climbed to another high point it would slow down as its KE transferred to GPE.

Figure 17.1

Kinetic energy

$$\text{KE} = \tfrac{1}{2}\,mv^2, \quad \text{KE} = \tfrac{1}{2}\,\text{mass} \times \text{speed}^2$$

Figure 17.2 shows a mass, **m**, at rest at a height, **h**, above the ground in a gravity field of strength, **g**.

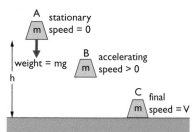

Figure 17.2

At A: all the energy is GPE = **mgh**

At B: the mass is falling so GPE is decreasing and KE is increasing. The total energy is the same but now total energy = GPE + KE

At C: the mass is about to hit the ground so now all the energy is KE = $\tfrac{1}{2}\,mv^2$

Note: GPE at the start = KE at the end

$$mgh = \tfrac{1}{2}\,mv^2$$

Worked Example 8

The diagram shows a roller coaster ride. The car is dragged to the highest point in the ride, A, and then released. The car, when loaded, has a mass of 800 kg. [Take g = 10 N/kg]

a) What is the GPE of the car at point A?

b) How fast is the car moving when it reaches point B? What assumptions have you made?

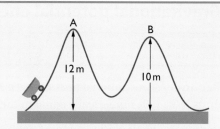

a) Use: GPE = *mgh*

So, GPE = 800 kg × 10 N/kg × 12 m

Therefore, GPE = 96 000 joules

b) Assuming total energy is conserved (none is 'lost' between A and B), then:

GPE at A = GPE at B + KE at B

So, KE at B = GPE at A - GPE at B → $\tfrac{1}{2}\,mv^2$ Therefore $\tfrac{1}{2}\,800\,v^2 = 800 \times 10 \times (12 - 10)$

So, $v^2 = 40$

Therefore, $v = \sqrt{40}$ = about 6.3 m/s

Note: remember that 'lost' here means converted to noise and heat.

Power

Power is the rate of transfer of energy, also expressed as the rate of doing work. You need to be able to calculate the power of a system using:

$$\text{power} = \frac{\text{work done}}{\text{time taken}} \qquad P = \frac{W}{t}$$

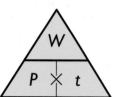

Power is measured in watts if work done is measured in joules and time is measured in seconds.

1 watt = 1 joule/second

Often questions will give you all you need to calculate *mechanical* work done and, given the time taken, th. power. You may meet questions in which you are asked to calculate the efficiency of a motor used to do mechanical work – remember that the power of an electric motor is given by $P = IV$.

Worked Example 9

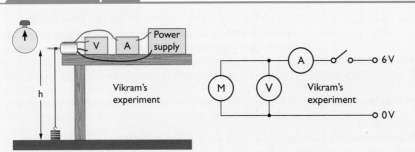

Vikram's experiment

Vikram's experiment

Vikram carried out an experiment to measure the efficiency of a small electric motor. He timed how long it took to raise a known load, **F**, through a measured height, **h**, so he could calculate the mechanical power output. He repeated the experiment to get an average value for **t**, the time taken to raise the load. He used an ammeter to measure the current, **I**, supplied to the motor and a voltmeter to measure **V**, the voltage across the motor.

Here are his results: Load, **F = 1 N**, **h = 0.5 m**, average **t = 3.5 s**, **I = 40 mA**, **V = 6 V**. Calculate the efficiency of Vikram's motor. Give your answer as a percentage.

$$\text{Efficiency} = \frac{\text{useful energy output from the system}}{\text{total energy output from the system}} \times 100\%$$

$$\text{so Efficiency} = \frac{\text{mechanical output power} \times t}{\text{electrical input power} \times t} \times 100\%$$

$$\text{mechanical output power} = \frac{F \times d}{t} \rightarrow \text{mechanical output power} = \frac{1 \times 0.5}{3.5} \rightarrow 0.142 \text{ W}$$

$$\text{electrical input power} = I \times V \rightarrow \text{electrical input power} = 0.04 \times 6 \rightarrow 0.240 \text{ W}$$

$$\text{so Efficiency} = \frac{0.142 \times t}{0.240 \times t} \times 100\% \rightarrow 59\%$$

Note that t cancels in the equation left.

Practical work

There is an experiment in the Student Book (page 148) that is simple to carry out. It involves doing work against gravity by running up stairs. It sometimes forms the basis of questions in exam papers. It should be understood that the measurement allows you to calculate the increase in GPE achieved by running up the stairs, but not the total energy output of the person under test – work will also be done against friction and a fair amount of heat will be generated by the exertion!

Chapter 18: Energy resources and electricity generation

You need to know about a number of energy resources and understand the energy transfers that take place in the generation of electricity.

The key process in the generation of electricity is turning mechanical energy into electricity, with a generator:

In many systems the kinetic energy is provided by a turbine driven by high pressure steam. To produce the steam, heat energy must be supplied and this can come from a variety of different sources:

nany sources of energy to provide the heat needed to boil water:

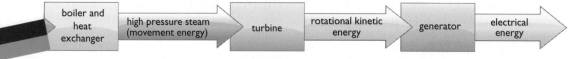

Geothermal energy is a naturally occurring energy produced by the heat of the Earth's core – this energy is accessible in places where volcanic activity is frequent. ✓

Solar heating is more frequently used for the direct heating of water, but with the use of focusing mirrors it is possible to heat water for high pressure steam production. ✓

Controlled nuclear fission can be used to produce near limitless amounts of heat. Although clean in terms of greenhouse gases there are problems with storing the radioactive waste securely and safely for very long periods of time. Nuclear fuel is a *non-renewable resource* ✖ (though it will not run out for centuries).

Fossil fuels include **oil**, **gas** and **coal**. These are *non-renewable* ✖ and produce greenhouse gases.

Hydroelectric power stations convert the GPE of water in mountain lakes and reservoirs to electricity by the following process:

Hydroelectricity is a *renewable* ✓ form of energy that produces no greenhouse gases. The energy comes ultimately from the Sun which evaporates water which then falls as rain in the mountains where it is collected.

Tidal power and wave energy: Barriers across tidal estuaries can harness the kinetic energy of the sea, which drives water turbines as the tide comes in. The water trapped behind the dam can then be used, as required, to drive turbines once the tide goes out. An increasing number of systems are being developed to collect the KE of waves. Ultimately the *renewable* ✓ energy source for both tidal and wave power is the Sun. The Sun (in conjunction with the Moon) causes tidal movement, and the combination of wind and tide generates waves.

Wind power: Windmills have been in use for thousands of years, converting the kinetic energy of the wind into other useful forms – for grinding flour, for powering machinery and for pumping water. More recently, wind farms using many wind driven generators have been built to convert the wind's kinetic energy into electricity. This *renewable* ✓ energy source has the Sun as its source. The effect of the Sun produces the pressure variations in our atmosphere that drive the wind.

Solar power: Solar power as a direct source of heat has been mentioned above. It is also used with solar heat exchangers to provide some of the hot water required for domestic use. Photovoltaic cells are also used to convert light from the Sun into electricity. This is a *renewable* ✓ source of energy that produces no CO_2 greenhouse gas. The amount of electrical energy generated by the use of photovoltaic cells is small.

Advantages and disadvantages

You will need to consider factors which influence the choice of energy used to generate electricity:

- **Cost**. This will include the cost of installing the power station and the cost of generating electricity. With nuclear power stations, a considerable cost element is that of safely decommissioning a reactor.

- **Renewable/non-renewable**. Our reserves of fossil fuels are finite so, in the future, alternatives must be found. Some of the finite resources, like petroleum, are vital ingredients of plastics, pharmaceuticals, etc.

- **Greenhouse gas emissions**. Burning fossil fuels produces CO_2 which is a greenhouse gas. Greenhouse gases trap heat in our atmosphere and contribute to global warming.

- **Supply and demand**. The demand on electricity varies over a period of 24 hours and longer. Nuclear reactors cannot be turned on and off quickly and therefore cannot respond to demand surges. Gas-fired power stations can bring extra generators into service to meet sudden demands. Some renewables, like wind power and tidal power, do not produce energy all the time.

- **Environmental impact**. Power stations can spoil the view. Waste products can damage plant and animal life in the vicinity of the power station. Even hot water waste can have adverse effects on rivers.

- **Location**. It makes sense to build power stations close to centres of demand, like urban and industrial areas, because energy will be wasted in transmission. Tidal, wind, geothermal and hydroelectric energy sources are limited to suitable geographical locations and these are often far away from centres of demand.

You will certainly be asked to consider advantages and disadvantages of different energy sources. It is not possible to produce an exhaustive list here. The above list should allow you to make valid points in an exam.

Worked Example 10

a) Photovoltaic cells have an efficiency of 15% (typically). If the total power reaching 1 m² of the Earth's surface is 600 W, what is the maximum electrical energy output you can expect from an array of photovoltaic cells covering 5 m²?

b) Give two reasons why the actual output power might be less than this.

a) The total input power from the Sun over an area of 5 m² = 600 W/m² × 5 m² = 3000 W

$$\text{efficiency} = \frac{\text{useful electrical power output}}{\text{total power output}} \times 100\%$$

Therefore, useful electrical power output = 3000 W × $\frac{15}{100}$ → 450 W

b) (There are many valid answers.) The intensity of the energy from the Sun will change during the day because of cloud cover and how high the Sun is in the sky. For maximum intensity the panels should be pointed towards the Sun. The efficiency of the panels may deteriorate over time; they may get dirty, reducing their efficiency.

Practical work

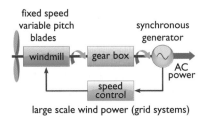

large scale wind power (grid systems)

Figure 18.1

Investigations can be carried out to find different systems for absorbing solar energy to directly heat water; this will use knowledge from the chapter on heat transfer. Simple wind generators can be made from DC motors and the power output can be measured.

Warm-up questions

1 Describe the energy transfers that take place when a trampolinist bounces up and down on a trampoline.

2 *a)* Explain what is meant by the law of conservation of energy.

b) Aquil says that if energy is conserved a ball in the figure overleaf, released at A, will roll down the slope and come to rest momentarily at B and continue to roll backwards and forwards between the two points indefinitely. Is he right or wrong? Explain your answer.

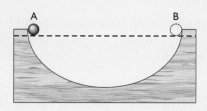

3 Explain the energy transformations that take place when a windmill on a wind farm is operating; start with the wind and finish with the output from the windmill.

4 A typical filament lamp is 8% efficient. What does this mean?

5 A lift motor is 70% efficient. It does 450 000 J of work lifting a load to the top floor of a building.

   a) What electrical energy input is required to do this?

   b) What unwanted energy transfers take place?

   c) Draw a labelled Sankey diagram to show the energy transfers.

6 State the three mechanisms by which heat may be transferred from one place to another.

7 State the main way in which heat energy is transferred in the following situations:

   a) From the Sun to the Earth.

   b) From a gas flame to the contents of a saucepan.

   c) From the heating element at the bottom of a kettle to the water at the top of the kettle.

8 During the day the land shown in the figure heats up quickly and its temperature will be greater than that of the sea. At night the land cools more quickly than the sea so its temperature is lower than that of the sea.

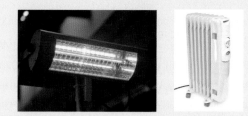

The result is a breeze which blows either from the land to the sea or from the sea to the land. State and explain which happens during the day and which happens at night.

9 A refrigerator uses a system that extracts heat from within an insulated cabinet and pumps it to the outside – and, as it does this, lowers the temperature inside the cabinet. The heat exchanger is shown below; what features of its design make it more efficient at losing heat to the surroundings?

10 Radiant heaters are often mounted overhead in shops and factories whereas convector heaters are mounted at ground level. See the figures below. Explain, in terms of heat transfer processes, why each is effective.

11 Explain how the following are effective at reducing heat transfer:

   a) Packaging take-away meals in shiny metal foil containers.

   b) Putting a plastic lid on take-away coffee cups, as shown below.

   c) Putting a ridged cardboard sleeve around the cup, as shown below.

   d) Fluffing up your feathers (if you are a bird)!

12 Building regulations have been changed over the last ten years to make new homes more energy efficient. Explain one way in which heat losses from:

a) the roof, b) the walls and c) the windows of houses can be reduced.

13 Why does wearing loose white clothes help to keep you cool in hot weather?

14 Although wetsuits used by scuba divers are made of rubber or neoprene they do not, as their name suggests, keep you dry. Does this make them more effective or less effective at keeping you warm when diving in cool water?

15 Calculate the work done in the following situations:

a) You lift a *weight* of 300 N from ground level to a height of 2.2 m.

b) You push a *car* 20 m along a level road at a steady speed against a friction force of 750 N.

c) Rocket motors make a *rocket* of mass 2200 kg accelerate at 2.5 m/s^2 over a distance of 650 m in deep space.

16 State how much energy has been transferred to the objects in italics in the previous question and the form of the energy gained.

a) The *weight* b) The *car* c) The *rocket*

17 The gravitational field strength, g, at the Earth's surface is 10 N/kg; on the Moon it is 1.7 N/kg.

a) How much gravitational potential energy (GPE) would be transferred to a mass of 30 kg if it was raised through a perpendicular height of 50 m on the Moon?

b) If the same amount of GPE was to be transferred to a mass of 500 grams on the Earth, how high would it be above the Earth's surface?

18 Put the following moving objects in order of the amount of kinetic energy they possess, from smallest to largest:

A: A 10 000 kg lorry travelling at 2 m/s.

B: A cheetah, mass 48 kg, running at 30 m/s.

C: A bullet of mass 4 grams travelling at 900 m/s.

D: An alpha particle, mass 7×10^{-27} kg, travelling at 30 000 km/s.

19 Amelia throws a ball up in the air and it eventually falls back to the ground. Put the following energy changes into the order that they occur, starting with Amelia throwing the ball and finishing when the ball is back on the ground:

A: The ball has only gravitational potential energy (GPE).

B: Amelia does work on the ball.

C: The ball has only kinetic energy (KE).

D: All the energy is converted to heat and sound.

E: GPE is being converted to KE as the force of gravity does work on the stone.

F: KE is being converted to GPE as the ball slows down.

G: The ball has only kinetic energy (KE).

20 a) Define the term power, giving an example involving power.

b) State the unit in which power is measured.

21 A roller coaster train weighs 12 000 N. It is lifted to the top of the ride through 30 m in 20 s.

a) Calculate the rate of transfer of energy from the electric motor to the train.

b) What energy conversions are taking place?

c) Give a reason why the transfer of energy from the electric motor to the train is not 100% efficient.

22 Give two examples of renewable energy sources.

23 What are fossil fuels? Give three examples of fossil fuels.

24 Tyrone does not understand how playing computer games can contribute to global warming (other than the fact that he breathes out carbon dioxide, but he does that all the time). Explain briefly why computer games can contribute to global warming.

25 Describe the energy transfers that take place in the generation of electricity at

a) A hydroelectric power station, and

b) a coal fired power station.

26 Give one advantage and one disadvantage of each of the following ways of generating electrical energy:

a) An oil fired power station.

b) A tidal power station.

c) A nuclear power station.

d) Wind farms.

e) Using photovoltaic cells to convert the Sun's energy directly to electricity.

Chapter 19: Density and pressure

Density

$$\text{density, } \rho = \frac{\text{mass, } m}{\text{volume, } V}$$

Density is measured in kg/m³, it is the mass of a 1 metre cube of a substance. A more convenient unit, g/cm³, is often used. To convert from g/cm³ to kg/m³ simply multiply by 1000.

For example, the density of water is 1 g/cm³ which is equivalent to 1000 kg/m³.

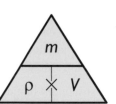

Measuring density

For a regular solid

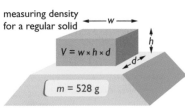

Figure 19.1

Measure the width, w, height, h, and depth, d, of the block in cm with a half metre rule.

Volume, $V = (w \times h \times d)$ cm³

Find the mass, m, of the block in grams using an electronic balance.

Use the formula to calculate the density.

For a liquid

Find the mass, m_1, of a dry, clean measuring cylinder. Pour in the liquid and find the new mass, m_2. Find the mass, m, of liquid using $m = m_2 - m_1$ grams. Alternatively, place the dry measuring cylinder on the balance and zero (tare) the balance so you can read off the mass of liquid directly when you re-weigh the cylinder.

Use the formula as above: Volume, $V = (w \times h \times d)$ cm³

Your measuring cylinder may be calibrated in ml (millilitres): $1 \ ml = 1 \ cm^3$

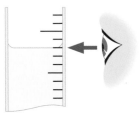

Figure 19.2

Read the volume, V, of liquid from the scale looking straight at the bottom of the curved surface of the liquid.

Worked Example 1

a) A block of wood that measures 20 cm × 5 cm × 10 cm has a mass of 528 g. Calculate the density of the wood in g/cm³ and in kg/m³.

b) 40 cm³ of sea water has a mass of 44.2 g. Calculate the density of the sea water.

a) density, $\rho = \dfrac{\text{mass, } m}{\text{volume, } V}$

So, $\rho = \dfrac{528 \text{ g}}{(20 \times 5 \times 10) \text{ cm}^3}$

Therefore , $\rho = 0.528$ g/cm³ or 528 kg/m³

b) density, ρ = mass, m / volume, V

So, ρ = 44.2 g / 40 cm³

Therefore, $\rho = 1.11$ g/cm³

Wood has a *lower* density than water, therefore it floats on water.

Sea water has a *higher* density than fresh water, so objects float more easily in sea water.

$$\text{pressure, } p = \frac{\text{force, } F}{\text{area, } A}$$

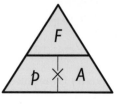

essure is measured in pascals (Pa). It is the weight, in newtons per metre², acting down on a surface.

The unit is $1 \text{ Pa} = 1 \text{ N/m}^2$.

Solid objects exert a downward pressure upon the surfaces on which they stand. Stiletto heels can exert a damagingly high pressure on surfaces because the person's weight is concentrated over a tiny area. Snow shoes and skis are designed to spread a person's weight over a larger area reducing the pressure on a soft snow surface.

In stationary liquids and gases, pressure at any point acts in all directions *and increases* with the depth of gas or liquid. A difference in pressure between two places in a liquid or gas will result in a flow of the liquid or gas from the place at higher pressure to the place at lower pressure.

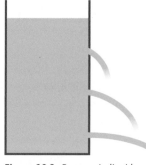

Figure 19.3 *Pressure in liquids increases with depth*

Air pressure exerts a force at sea level of around 100 000 N on each square metre; this pressure, 100 kPa (kilopascals) is equivalent to 10 tonnes on each square metre on the ground, walls and us!

Worked Example 2

A clean empty oil can is shown above. Although made of thin bendable metal it keeps its shape. When all the air is pumped out of it, as shown above, right, the can is crushed. Explain each situation.

When there is air inside the can it pushes outwards on the can with exactly the same pressure as the air outside is pushing inwards. As the pressure is balanced inside and outside the can keeps its shape. When the air is pumped out the pressure inside the can pushing outwards falls to a very low value. As the air pressure pushing in on the can is now no longer balanced the can is crushed.

Pressure and depth

pressure difference, *p*, in Pa = height in m × density in kg/m³ × gravitational field strength in N/kg

$$p = h \times \rho \times g$$

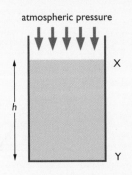

atmospheric pressure

What is the pressure difference between point **X**, at the top surface of water in a tank, and at point **Y**, at the bottom of the tank? The density, ρ, of the water is 1000 kg/m³, the height, **h**, of water in the tank is 5 m and the cross-sectional area of the tank, **A**, is 4.8 m². Show how you reach your answer.

The volume of water in the tank is **h** × **A** therefore the mass of water in the tank is **h** × **A** × ρ

So, the weight force of water acting down is **h** × **A** × ρ × **g**.

Use: pressure = force/area, giving $p = h \times A \times \rho \times \frac{g}{A}$

Therefore, cancelling out A, the **pressure difference** between X and Y is (5 m × 1000 kg/m³ × 9.8 N/kg)

Therefore, the answer is 49 000 N/m²

(i) The cross-sectional area of the tank makes no difference to your answer – you can see it cancels out in the pressure equation.

(ii) The total pressure on the bottom of the tank is the pressure at X, the top of the tank (due to atmospheric pressure), *plus* the pressure difference you have calculated. If atmospheric pressure is 100 000 N/m² then the total pressure at the bottom of the tank is 149 000 N/m² or 149 000 Pa.

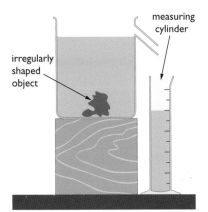

Figure 19.4

Practical work

You should be able to find the density of an irregularly shaped solid. Weigh the object to find its mass as usual. Now you must find the object's volume. The apparatus needed to do this is shown in Figure 19.4. First fill a can with a spout with water until it just flows out of the spout. Place a clean, dry measuring cylinder beneath the spout and then slowly lower the irregular solid into the can until it is completely submerged. The water displaced will have the same volume as that of the object, so the volume of the object is found by measuring the volume of water collected in the measuring cylinder.

The density can then be calculated in the usual way.

Chapter 20: Solids, liquids and gases

Please see the Appendix for some additional material regarding changes of state.

The states of matter

Matter can exist as a solid, liquid or gas. Heating and cooling can cause changes of state as shown below:

Figure 20.1

The changes of state in this diagram occur at specific temperatures: solid ⇔ liquid at the freezing/melting point of the substance and liquid ⇔ gas at the boiling point of the substance.

Evaporation is the change of state from liquid to gas that occurs at temperatures below boiling point; liquids evaporate more slowly at lower temperatures.

The molecules in solids are tightly packed and held in fixed positions by strong forces. The molecules vibrate around their fixed positions. As the solid gets hotter these vibrations get bigger.

In a liquid the molecules are closely packed but do not have a regular structure. The forces between the molecules are strong. The molecules move randomly.

The molecules in a gas are widely spaced and in a continuous state of random motion. The forces between the molecules are very small except during collisions. The molecules move randomly.

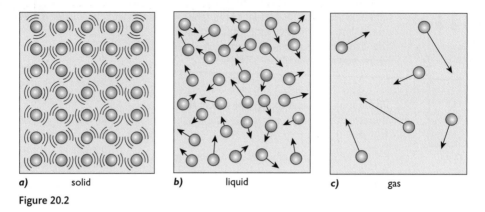

a)   solid   *b)*   liquid   *c)*   gas

Figure 20.2

> **Worked Example 4**

a) Water can exist as a solid (ice), a liquid, and a gas (water vapour). Explain why its density as a liquid is similar to its density as a solid, whilst its density as a gas is very much smaller.

b) Convection can take place in liquids and gases but not in solids. Explain this in terms of the molecular structure of each state of matter.

c) In which state(s) can water be compressed easily? Explain why it can be compressed easily in these state(s) and not in its other state(s).

a) The molecules are closely packed in both water and ice; in the gas state the molecules are widely separated. Therefore the mass per unit volume in water vapour is much lower.

b) The molecules in solids are in fixed positions and cannot move randomly. In both gases and liquids the molecules are not fixed in any position and move relative to one another – this allows for circulation of the liquid or the gas due to local heating.

c) Water (and all substances) can be compressed in the gas state because the molecules are widely separated. In solids and liquids the molecules are close together and exert large forces on each other.

Brownian motion

If you look at tiny particles of matter, like smoke particles, in air under a microscope they are continuously jiggling around as if they are being struck randomly on all sides by invisible particles. The scientist Robert Brown was the first to report this effect which he observed whilst looking at tiny pollen grains in liquid under a microscope.

Therefore, the effect, **Brownian motion**, is named after him.

It provided evidence for our model of the structure of matter, called the kinetic theory of matter.

Figure 20.3

The particles in matter (molecules) are extremely small and, in liquids and gases, they are in a continuous state of rapid random motion.

A gas exerts a pressure on objects, for example, on the walls of a container as a result of the continuous collisions between the gas molecules and the container. Each individual collision produces only a tiny force but the collisions of many millions of molecules colliding per second result in a significant pressure. Here are some key points you should know:

- The random motion of gas and liquid particles explains why pressure acts in all directions at any point.
- The speed of molecules increases with temperature, so as we heat gases in a rigid container, more energetic collisions with the walls occur more frequently, and the pressure of the gas increases.
- The temperature of a gas in kelvin is proportional to the average kinetic energy of the gas molecules.

The gas laws

The gas laws refer to the behaviour of fixed amounts of gas. You should know how the following properties of a fixed amount of gas change with respect to one another: **VOLUME, V, PRESSURE, p, and TEMPERATURE, T.**

Boyle's Law

For a fixed amount of gas at constant temperature

$$p_1 \times V_1 = p_2 \times V_2$$

Or, rearranged,

$$\frac{p_1}{p_2} = \frac{V_2}{V_1}$$

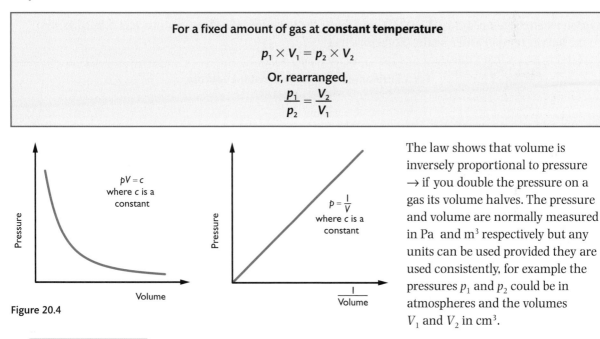

Figure 20.4

The law shows that volume is inversely proportional to pressure → if you double the pressure on a gas its volume halves. The pressure and volume are normally measured in Pa and m^3 respectively but any units can be used provided they are used consistently, for example the pressures p_1 and p_2 could be in atmospheres and the volumes V_1 and V_2 in cm^3.

Worked Example 5

A syringe is filled with 20 cm³ of air at atmospheric pressure (100 kPa). Benedict puts his thumb over the end of the syringe so no air can enter or leave the syringe and then pushes the plunger of the syringe until the air is compressed to a volume of 12 cm³. What is now the pressure of the air inside the syringe? What happens when the plunger is released?

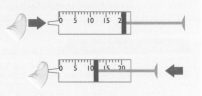

Rearrange $\frac{p_1}{p_2} = \frac{V_2}{V_1}$ to give $p_2 = p_1 \times \frac{V_1}{V_2}$.

So, $p_2 = 100 \text{ kPa} \times \dfrac{20 \text{ cm}^3}{12 \text{ cm}^3}$

Therefore the new pressure, $\boldsymbol{p_2} = 166.7$ kPa

If the plunger is released with the thumb still trapping the air it will spring back to its original volume and pressure. The movement results from the pressure difference between the compressed air inside and the atmosphere outside.

olute zero

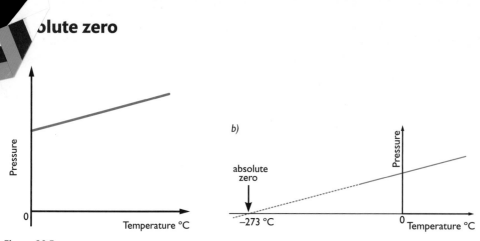

Figure 20.5

If you cool a fixed mass of gas at constant volume its pressure drops as shown in graph a. If you continue to cool the gas below 0°C the trend continues with the pressure dropping at a uniform rate. If the graph is extended the pressure will keep on dropping until the pressure is zero – the graph predicts that this will happen when the temperature is −273°C and this is the lowest possible temperature. It is defined as zero on the **kelvin temperature scale**, 0K (zero kelvin).

For a fixed amount of gas at constant volume

$$\frac{p_1}{T_1} = \frac{p_2}{T_2}$$

or pressure is proportional to absolute (kelvin) temperature

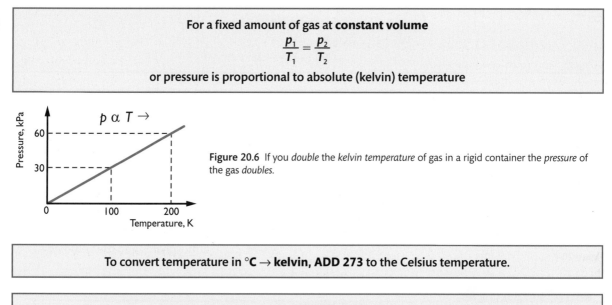

Figure 20.6 If you *double* the *kelvin temperature* of gas in a rigid container the *pressure* of the gas *doubles*.

To convert temperature in °C → kelvin, ADD 273 to the Celsius temperature.

To convert temperature in kelvin → °C, SUBTRACT 273 from the kelvin temperature.

Worked Example 6

1 Convert the following Celsius scale temperatures to kelvin scale temperatures:

   a) 20°C   *b)* 0°C   *c)* 100°C   *d)* 37°C   *e)* −33°C

2 Convert the following kelvin scale temperatures to Celsius scale temperatures:

   a) 280K   *b)* 233K   *c)* 300K   *d)* 23K   *e)* 345K.

3 To what temperature would you need to heat a fixed amount of air in a rigid container, initially at room temperature of 20°C, to cause the pressure inside the container to treble? Give your answer in °C.

1 In each case, add 273 to the Celsius temperature,

    e.g. *a)* 20 + 273 = 293K   *b)* 273K   *c)* 373K   *d)* 310K   *e)* 240K

2 In each case, subtract 273 from the kelvin temperature,

    e.g. *a)* 280 − 273 = 7°C   *b)* −40°C   c) 27°C   *d)* −250°C   *e)* 72°C

3 Let the initial pressure be p_1 (you can make up an arbitrary value, say 100 kPa if you prefer); the initial temperature of 20°C *must be converted to kelvin* → $T_1 = 293$K. The new pressure at T_2 must be $3p_1$ (or 300 kPa). Now use the formula

$$\frac{p_1}{T_1} = \frac{p_2}{T_2}$$

So, $\dfrac{p_1}{293\text{K}} = \dfrac{3p_1}{T_2}$

Cancel p_1 (common term) and rearrange to give:

$T_2 = 3T_1 = 3 \times 293\text{K} = 879\text{K}.$

Now convert to Celsius by subtracting 273 to get

$T_2 = 879 - 273$

$T_2 = 606°C$

Practical work

If possible do the simple experiment to observe the Brownian motion of smoke particles in a smoke cell. This requires a low power microscope, a tiny smoke 'cell' to hold some smoke and a bright light shining onto the smoke from the side. Otherwise, 'Brownian motion' on an internet search engine will produce a large number of video clips and computer animations.

The two key equations in the specification: $p_1 \times V_1 = p_2 \times V_2$ and $p_1/T_1 = p_2/T_2$ can be verified by straightforward experiments using school/college apparatus. The key assumptions are that temperature remains constant during the pressure volume (Boyle's Law) experiment and that volume remains constant when investigating the pressure temperature relationship.

The relationships are true for 'ideal' gases – ideal gases cannot be liquefied. Ideal gases don't exist of course, but real gases behave well enough like ideal gases provided they are not put under very high pressures or cooled to very low temperatures, neither of which is likely to happen in a school laboratory.

You are likely to get poor results if moist air is used; water vapour is an example of a gas that is far from ideal because it will readily convert to a liquid as its temperature drops.

Warm-up questions

1 Copy and complete the following statements about density:

Density is a measure of how tightly matter is packed within a particular substance. To calculate the density of the substance an object is made of you must divide the _____ of the object by the _____ of the object. The units of density are _____ per _____ or _____ per _____ .

2 Describe how you would find the density of the following:

    a) a rectangular block of glass,

    b) a steel ball bearing, and

    c) a stone.

You should explain the things you need to measure, how you are going to measure them and how you will use the measurements to calculate the density in each case.

3 a) State the formula for calculating the pressure exerted on a surface by matter, whether it is solid, liquid or gas.

b) What is the unit of pressure?

4 A rectangular block of metal measures 10 cm by 8 cm by 5 cm and has a mass of 2.6 kg.

a) How much does it weigh? [Take $g = 10$ m/s^2]

b) Calculate the maximum and minimum pressure it will exert on a surface, explaining why there is a difference.

5 Look at the two U-tubes filled with water shown in the figure below. They represent a freeze frame in a video of an experiment.

a) Assuming both tubes are open to the atmosphere, what can you say about the pressure acting on either side of the point X in each diagram?

b) What would you expect to see happen in each case if the video were allowed to run on?

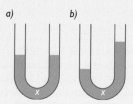

a) b)

6 'The pressure at any point in a gas or a liquid acts equally in all directions.' Explain why this is so.

7 The figure shows a simple mercury barometer. It consists of a strong glass tube filled with mercury with its open end in a small reservoir of mercury. The column of mercury is supported by the pressure of the air acting down on the surface of the mercury in the reservoir.

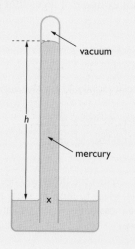

vacuum

h

mercury

x

a) If the atmospheric pressure supports a column of mercury of height, **h**, 0.76 m, what is the atmospheric pressure in standard units if g is 9.8 m/s^2 (or 9.8 N/kg) and the density of mercury is 13 600 kg/m^3?

b) What would happen if the atmospheric pressure dropped by a small amount?

c) Why do you think mercury is used instead of water in barometers of this type?

8 Copy and complete the following statements about change of state of matter.

When a solid substance is heated to its _____ point it starts to turn into a _____; when it has all melted further heating will raise the temperature speeding up the process called _____ by which changes state to a _____. When the temperature of the substance reaches its _____ point the temperature remains constant until all of the substance has changed state from _____ to _____.

9 What is the difference between evaporation and boiling?

10 Copy and complete the table below, which summarises the properties of solids, liquids and gases; some sections have been completed as an example.

Property	Solids	Liquids	Gases
Has a definite shape			No
Easily compressed	No		
Density		High	
Can be poured (fluid)		Yes	
Expands to fill all available space	No		

11 Draw sketches to show the arrangement and motion of the particles that make up a solid, a liquid and a gas. Your sketches should be labelled to highlight the key features of each state of matter.

12 In an experiment to demonstrate Brownian motion smoke particles observed under a microscope were seen to be jiggling around continuously. State two main deductions made about the cause of this motion.

13 The behaviour of gases is explained in terms of the movement of the molecules of the gas, a theory called the kinetic theory of gases.

a) How does this explain the pressure exerted by a gas on the walls of a container that it is in?

b) How does it explain the fact that the pressure exerted on the walls of a container of fixed volume increases with temperature?

c) If we cool a gas, what happens to the way in which the molecules move?

d) Is there any limit to how cold a gas can be made and how does this relate to the theory?

14 Copy and complete the graph, which shows the relationship between the pressure of a fixed volume of gas and its temperature in degrees Celsius. Label the axes and mark important points on the horizontal scale.

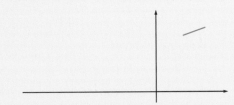

15 If the temperature of a gas is increased from –23°C to 227°C what happens to the average kinetic energy of the gas molecules?

16 The required pressure of the air in a car's tyres is specified in the owner's handbook. If you check your tyre pressure on a very hot summer's day and find it is just at the required level, what will happen if there is a significant drop in temperature a few weeks later? [Assume that there is very little change in the tyre volume.]

17 State the relationship between the pressure p_1 of a fixed volume of gas at temperature T_1 and its new pressure p_2 when at the same volume when the temperature is T_2.

18 A rigid canister is filled with gas at a pressure of 200 kPa at a temperature of 10°C. If the canister can withstand a maximum pressure of 300 kPa, what is the maximum temperature to which it can be safely raised?

19 [Harder question] The figure shows a constant volume gas thermometer. A fixed mass of gas is trapped inside the round bulb by mercury in a U-tube, as shown. Atmospheric pressure acts on the open side of the U-tube and is 100 kPa (you can take this as equivalent to 76 cm of mercury). Initially the trapped gas is at 0°C and the levels of the mercury in both sides of the U-tube are the same.

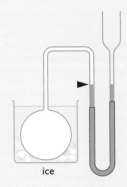

ice

The constant volume mark is indicated by the arrow. The air in the bulb is now heated to 100°C.

a) What will initially happen to the mercury level in the the two sides of the U-tube? Give your reason.

b) Mercury is added to the open side of the U-tube until the mercury level in the left-hand side of the U-tube is restored to the constant volume mark. What height will the mercury in the right-hand side of the U-tube now be, above the constant volume mark in the left-hand side of the U-tube?

20 A balloon is filled with air at sea level and taken up to a height of 3000 m in an unpressurised aircraft. Describe and explain what you expect to see happening to the balloon as the plane flew up to this height. (Assume that, somehow, the temperature around the balloon remains constant!)

21 A diving bell is lowered into the sea, as shown below. Describe and explain what you would expect to happen to the space containing trapped air as the bell went lower and lower into the sea.

(Assume that the air inside stayed at a constant temperature.)

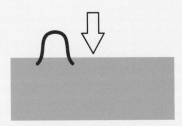

22 a) State the relationship between the pressure p_1 of a fixed amount of gas with a volume V_1 and its new pressure p_2 when the volume is changed to V_2 without change in temperature.

b) The volume of gas trapped by a piston is increased from 40 cm³ to 200 cm³ without change in temperature. If the new pressure of the gas is 25 kPa what was the original pressure of the gas? See the figure on the right.

c) *[Harder].* Assuming that atmospheric pressure is 100 kPa and that there is very little friction between the piston and the cylinder, what would you expect to happen when the force holding the piston in either the starting position or the finishing position is removed? Show any calculations.

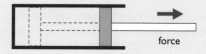

force

Notes

Chapter 21: Magnetism and electromagnetism

Magnets and magnetic materials

Certain materials, notably iron and iron compounds, display the effect we call magnetism. This is the ability to attract and pick up other magnetic substances. Unmagnetised iron can be magnetised and demagnetised quite easily, but permanent magnets are usually made of steel or a range of modern magnetically 'hard' materials.

Hard magnetic materials are used for things like magnets that we want to stay magnetised. **Soft** magnetic materials are used when we need something to magnetise and demagnetise easily.

Figure 21.1

Worked Example 1

The photographs above show a number of applications of magnetic materials. State which should be made of a *soft* magnetic material and which should be made of a *hard* magnetic material.

The compass needle and the fridge magnets need to keep their magnetic effect, so a magnetically *hard* material should be used.

The electromagnet consists of an iron disc which is strongly magnetised by an electric current in a coil, but it loses most of its magnetism when the current is turned off, so that it can release the scrap iron. It should be made from a *soft* magnetic material.

The transformer core needs to magnetised and remagnetised by an alternating current so its core should be made from a *soft* magnetic material.

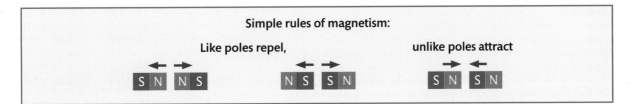

Simple rules of magnetism:

Like poles repel,   unlike poles attract

Unmagnetised iron, steel and a few other elements and their alloys or compounds can be attracted to a magnet. The permanent magnet induces magnetic effect in the unmagnetised iron with an unlike pole nearest the pole of the magnet. Therefore unmagnetised materials like iron are always attracted to magnets, never repelled.

Magnetic field lines

When iron filings are sprinkled onto a sheet of paper placed over a bar magnet they become temporarily magnetised as described above. As they each behave like tiny bar magnets they form lines with North and

South poles attracting each other. This suggests the idea of lines of magnetic force and the idea has become one way of describing the magnetic field, although the field is continuous around a magnet.

> The lines show the *direction of the force that would act on a small N pole* placed at any point in the field, so the arrows on the lines always point away from the N pole towards the S pole.

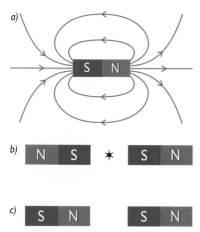

a)

The direction of the force at any point in a magnetic field is unique (or zero) so *magnetic field lines do not cross*.

The *field lines* are drawn *closest together where the magnetic field is strongest* and are drawn further and further apart as the field gets weaker. The magnetic field pattern for a single bar magnet (Figure 21.2a) shows that the strongest parts of the magnetic field are close to the poles.

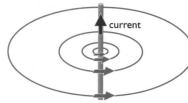

b)

You should also be able to complete the field patterns for the arrangements of magnets shown in Figure 21.2. In Figure 21.2b, where two like poles are facing, note the *neutral point*,⊙, midway between the two poles. In Figure 21.2c the field is uniform between the facing N and S poles.

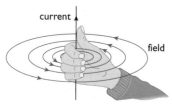

c)

Figure 21.2 *The field between unlike magnetic poles is uniform*

Electromagnetism

When a current flows in a wire a magnetic field is produced. Figure 21.3 shows the circular magnetic field produced by a steady current flowing in a long straight wire. **Remember** that the field is continuous (and extends over the whole length of the wire); the field lines represent the direction of the force a N pole would feel and the spacing of the field lines shows that the strength of the field decreases with distance from the wire. Increasing the current will increase the strength of the magnetic field produced.

The direction of the field line arrows can be remembered by the *right-hand grip rule*, as shown in Figure 21.4.

You should also be able to sketch the magnetic field patterns shown in Figures 21.5 and 21.6.

Figure 21.3

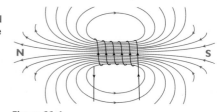

Figure 21.4

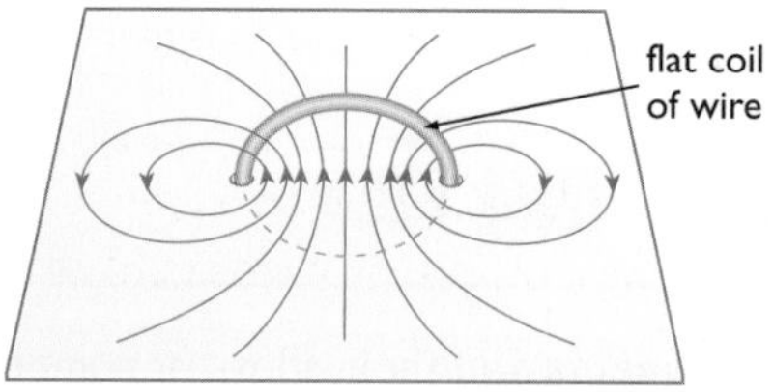

Figure 21.5

Figure 21.6

Electromagnets

You have seen that a current in a coil of wire produces a magnetic field. This field can be made stronger by increasing the current through the coil, increasing the number of turns on the coil or by winding the coil around a magnetic material. A 'soft' material is used for this purpose – the core should magnetise easily when the current flows through the coil but must demagnetise when the current is turned off.

You should be able to describe how electromagnets are used in simple electric bells, electromagnetic relay switches and solenoids. A solenoid is used to open a door latch remotely. They are used in blocks of flats to allow guests into the block when a resident, from his flat, pushes a switch to release the latch of the door to the block.

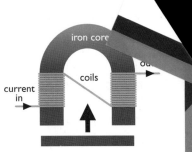

Figure 21.7 *The iron plate is attracted to the* **electromagnet** *when a current is flowing through the coils around the iron core.*

Practical work

You should know how to show magnetic field patterns produced by simple arrangements of bar magnets using iron filings. Very powerful rare earth magnets are now widely available, but this experiment works better with the weaker steel bar magnets. You should also be able to plot magnetic fields using small plotting compasses. Both methods are shown in Figure 21.8, though it is customary to place the magnet under the paper to prevent contact with the filings.

Plotting compasses are the best way of showing the shape of magnetic fields produced by a long straight wire, a flat coil and a long coil (solenoid).

Figure 21.8

You should be able to describe and explain a simple experiment to show how the strength of an electromagnet depends on current, number of turns on the coil and (possibly) that only certain materials inside the coil make good electromagnets. A suitable experimental set-up is shown in Figure 21.9. When winding your coils it is simpler to just wind a single coil at the top of your iron core. U-shaped cores will produce a *much* more powerful electromagnet, but the experiment can be carried out with a rectangular strip of iron or even an iron nail.

The wire must be insulated so the coils do not short together. **You must be careful about how much current you pass through the coil;** it can overheat which will melt plastic insulation and could cause a serious burn.

If you cannot wind your own coils with different numbers of turns you can still change the current through the coil, but again take care not to exceed the maximum safe current for the coil.

You should be able to present your results in graphical form and, for the exam, read and interpret the graphical results of experiments such as this.

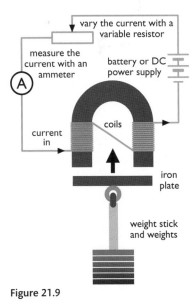

Figure 21.9

Chapter 22: Electric motors and electromagnetic induction

The motor effect – movement from electricity

When a current is passed through a wire placed in a magnetic field a force is produced which acts on the wire.

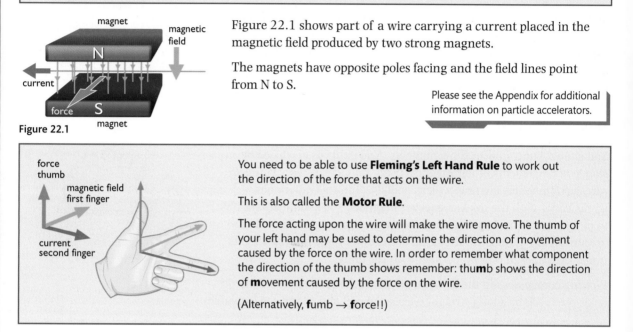

Figure 22.1

Figure 22.1 shows part of a wire carrying a current placed in the magnetic field produced by two strong magnets.

The magnets have opposite poles facing and the field lines point from N to S.

Please see the Appendix for additional information on particle accelerators.

You need to be able to use **Fleming's Left Hand Rule** to work out the direction of the force that acts on the wire.

This is also called the **Motor Rule**.

The force acting upon the wire will make the wire move. The thumb of your left hand may be used to determine the direction of movement caused by the force on the wire. In order to remember what component the direction of the thumb shows remember: thu**m**b shows the direction of **m**ovement caused by the force on the wire.

(Alternatively, **f**umb → **f**orce!!)

Moving coil loudspeaker

The motor effect is also used in **loudspeakers**: the signal current produced by an amplifier is alternating, and by passing it through a coil in a magnetic field the current results in alternating forces on the coil. The coil is attached to a paper cone and this transfers the vibrations to the air.

Worked Example 2

a) Copy and complete the following diagrams by drawing in magnetic field lines.

b) Use Fleming's Left Hand Rule to work out the direction of the force that will act on the conductors shown in the magnetic fields, below.

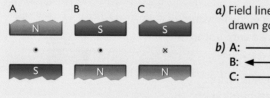

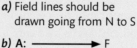

a) Field lines should be drawn going from N to S

b) A: ⟶ F
   B: ⟵ F
   C: ⟶ F

● Shows a conductor carrying current flowing **out of** the plane of the paper perpendicularly **towards** you.

✗ Shows a conductor carrying current flowing **into** the plane of the paper perpendicularly **away from** you.

The *size of the force* that acts on a current-carrying conductor placed at right angles to a magnetic field may be increased by either *increasing the strength of the magnetic field* or by *increasing the current* in the wire.

The electric motor

In its simplest form a DC motor consists of a single turn coil of wire that is free to rotate in a magnetic field about an axle, as shown in Figure 22.2. Carbon brushes make contact with the ends of the coil that are connected to a commutator so that a current can be passed through the coil.

The sequence of diagrams in Figure 22.2 show the coil from an end-on view, making it easy to see how the forces acting on each side of the coil produce a turning effect about the axle. Figure 22.2c shows that the turning effect is zero when the coil is parallel to the permanent magnets (because the line of action of the forces

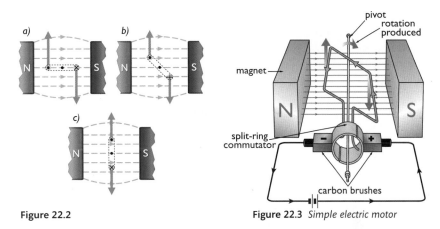

Figure 22.2

Figure 22.3 *Simple electric motor*

passes through the axis of rotation). This might suggest that the coil stops in this position, but it will inevitably overshoot, and as soon as it does so, the commutator will reverse the direction of the current in the coil which means the coil will continue to spin.

Worked Example 3

State three ways in which you could change the design of a DC motor to make it spin faster for a given load.

Increase the strength of the magnetic field. Put more turns on the coil. Pass a larger current through the coil. (But note that if the maximum design current for a motor is exceeded then the motor is likely to burn out.)

Electromagnetic induction and the generator

When a conductor is in a changing magnetic field a voltage will be induced in the conductor.

The magnetic field can change if:

- the conductor is moving into, or out of, a magnetic field (see Figure 22.4a),
- a magnet is moving towards, or away from, the conductor (see Figure 22.4b) or
- the magnetic field is being varied (see Figure 22.4c).

If the conductor is part of a closed electric circuit then the induced voltage will cause a current to flow.

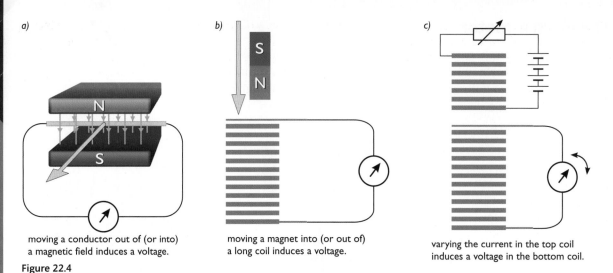

a)

moving a conductor out of (or into) a magnetic field induces a voltage.

b)

moving a magnet into (or out of) a long coil induces a voltage.

c)

varying the current in the top coil induces a voltage in the bottom coil.

Figure 22.4

> The size of the induced voltage in a coil can be increased by increasing the rate of change of the strength of the magnetic field, by having more turns on the coil and by having a coil of greater area.

Worked Example 4

Look at the three situations shown below; in each case a small plane coil is in the field between two permanent magnets:

A   B   C

In **A** the coil is moving horizontally at constant speed between the two magnets.

In **B** the coil is moving horizontally at constant speed out of the gap between the two magnets.

In **C** the coil is stationary between the two magnets.

In which situation, if any, is a voltage induced in the coil?

In **A** the coil is moving in a uniform magnetic field so the amount of magnetic flux cutting the coil is not changing → No induced voltage

In **B** the coil is moving out of a uniform magnetic field so the amount of magnetic flux cutting the coil is decreasing → Voltage induced

In **C** the coil is not moving so the magnetic flux cutting the coil is constant → No induced voltage

Two types of generator

The *bicycle dynamo* consists of a strong, small bar magnet which is spun by the wheel of the bicycle. It spins between the faces of a U-shaped iron core on which a coil of wire is wound. The changing magnetic field induces an alternating voltage in the coil.

Larger *generators* have spinning coils of wire in the magnetic field of strong permanent magnets (or, in la...
generators, a magnetic field produced by electromagnets). Since the coil in which the voltage is being
induced is spinning, the electrical connections are made by brushes which slide over slip rings.

The transformer

> The function of a transformer is to change the size of an alternating voltage. This is done by having two separate
> coils with different numbers of turns.

Transformers consist of a core made from thin sheets of a
magnetically soft material clamped together. Two separate coils of
wire, insulated from one another, are tightly wound onto the core.
Transformers are designed to perform the job of changing voltage
with very little power loss → you may **assume that they are 100%**
efficient.

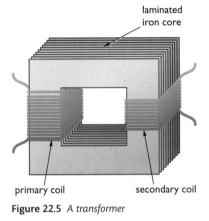

laminated
iron core

input (primary) voltage	=	primary turns
output (secondary) voltage		secondary turns

$$\frac{Vp}{Vs} = \frac{n_1}{n_2}$$

primary coil   secondary coil

Figure 22.5 *A transformer*

If $n_1 > n_2$ then the transformer *steps down* the input voltage; if $n_2 > n_1$ then the transformer *steps up* the
input voltage.

Worked Example 5

A transformer is designed to step down the mains voltage of 230 V to 11.5 V. If there are 1200 turns on the primary
coil how many turns should be wound on the secondary coil?

Rearrange $\frac{Vp}{Vs} = \frac{n_1}{n_2}$ to give $n_2 = \frac{Vs}{Vp} \times n_1$

So, $n_2 = \frac{11.5\ V}{230\ V} \times 1200$

Therefore, $n_2 = 60$ turns

Transmission of electrical energy

Transformers are used in the transmission of electrical energy over large distances. Transmission lines
have low but not zero resistance. Power loss due to this resistance is given by the formula $P = I^2R$, and this
means that power losses between the power station and the consumers would be unacceptably large. As
transformers are close to 100% efficient

Power input = power output   so $Vp \times Ip = Vs \times Is$ → $\frac{Vp}{Vs} = \frac{Is}{Ip}$ $\left(= \frac{n_1}{n_2} \right)$

So if a transformer is used to step up the generated alternating voltage 50 times this means the current
is stepped down 50 times → $Is = 1/50\ Ip$ The advantage of this is clear, since power loss along the
transmission lines is proportional to current squared this will reduce the loss by $(1/50)^2$. The voltage is
stepped down using transformers close to the consumers, again with very little power loss because of the
near 100% efficiency of the transformer.

...al work

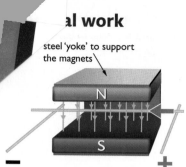

steel 'yoke' to support the magnets

N

S

—   +

Figure 22.6

You should have practical experience of both the motor and generator effects. A standard experiment to demonstrate that a current in a wire, placed at right angles to a magnetic field, produces a force on the wire is shown here. A short length of copper wire rests on two copper wire 'rails'. When a current is passed through the wire it catapults out of the field sliding along the rails. (The motor effect.)

The generator effect can be demonstrated by thrusting a strong magnet into a long, tightly wound coil connected to a galvanometer (sensitive ammeter). The induced voltage will circulate a detectable current. It can be seen that the induced voltage only occurs when the magnet is moving with respect to the coil. We can also observe that the direction of the induced voltage (and current) depends on the pole of the magnet entering the coil and its direction of travel. Faster movement induces a larger voltage.

Warm-up questions

1   A compass needle points north as shown in a) below.

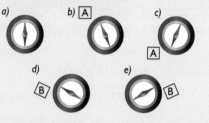

a)   b) [A]   c)

[A]

d)   e)

[B]   [B]

An unknown object **A** is brought near to the compass and the compass deflects towards it as shown in b) and c). A second unknown object **B** is brought close to the compass and it responds as shown in d) and e). What does this tell you about objects **A** and **B**?

2   Steel is a magnetically hard substance and it is often used for making compass needles. What is meant by the term 'hard' in this context?

3   The magnetic field around a magnet is often represented using *field lines*.

a)   What does the spacing of the field lines tell you about the magnetic field?

b)   What do the arrows on the field lines tell you about the magnetic field?

4   When a steel paper clip is suspended from a permanent magnet it is then able to pick up a further paper clip and this may be repeated several times to form a chain of paper clips hanging from the magnet. The paper clips have temporary or _____ magnetism. [Supply the missing word.]

5   **a)**   Show, with the aid of a field line sketch, how the magnetic field varies around a single bar magnet.

b)   Show how you would position two bar magnets to produce a region of strong and nearly uniform magnetic field. Add field lines to show the magnetic field.

6   Copy and complete the following magnetic field patterns for the three arrangements of current-carrying conductors shown.

⊗   *A long straight conductor carrying a current into the plane of the paper at 90°.*

⊙   ⊗   *A section through a plane circular coil carrying a current up out of plane of the paper on the left and down into the plane of the paper on the right.*

   A long solenoid wound on a cardboard tube. Current flowing in and out as shown by the arrows.

7   State two ways in which the strength of the magnetic field produced by the solenoid in question 6 could be increased.

8   A uniform magnetic field is used in cloud chambers to distinguish between different particles produced by collisions between subatomic particles. It is possible to tell the difference between positively and negatively charged moving particles – how do their paths differ, provided they are not moving parallel to the magnetic field lines?

9   Copy and complete the following paragraph:

Fleming's Left Hand Rule is used to predict the direction of the _____ on a _____ carrying conductor in a _____ _____. The thumb, first finger and second finger are arranged to point in three mutually perpendicular directions; the first finger points in the direction of the _____ _____, the second finger points in the direction of the _____ and the thumb indicated the direction of the _____ on the conductor. Fleming's Left Hand Rule is also known as the _____ rule.

10 Imagine that there is a uniform magnetic field acting over the area of this page directed perpendicularly into the page shown below. A current flows through a wire placed in the field.

  a) Copy the drawing and use Fleming's Left Hand Rule to find the direction of the force that acts on the conductor and mark it clearly on your diagram.

  b) State two ways in which the force on the conductor could be made weaker.

11 In which of the following will there be a voltage induced in the systems of conductors shown?

A bar magnet moving into a coil.

A bar magnet moving out of a coil.

A bar magnet moving whilst completely inside a coil.

A bar magnet at rest inside a coil.

12 Show, with the aid of a clear, labelled diagram, the key features of a transformer.

13 A transformer is designed to step a voltage down from 120 V to 6 V. Explain how the design you have shown in question 12 can do this.

14 What is the advantage of stepping up the voltage of electricity generated at a power station to a much higher voltage before transmitting electrical energy via the National Grid?

15 Transformers are assumed to be virtually 100% efficient. What does this mean in terms of electrical power input and output?

16 A generator consists of a coil which is driven mechanically to rotate in a magnetic field. How do the following factors affect the size of the voltage induced in the coil (if at all):

  a) The coil has more turns.

  b) The coil rotates faster.

  c) The magnetic field is made stronger.

Section F: Magnetism and Electromagnetism

Notes

Chapter 23: Atoms and radioactivity

Electrons, protons and neutrons

Atoms are made up of electrons, protons and neutrons. For example, here is a diagrammatic structure for an atom of the most abundant form of carbon, carbon-12.

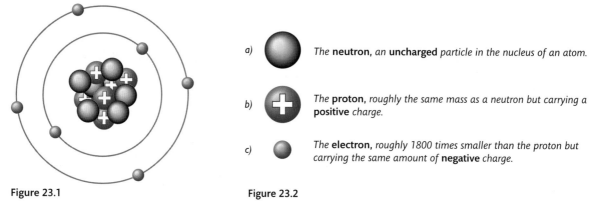

a) The **neutron**, an **uncharged** particle in the nucleus of an atom.

b) The **proton**, roughly the same mass as a neutron but carrying a **positive** charge.

c) The **electron**, roughly 1800 times smaller than the proton but carrying the same amount of **negative** charge.

Figure 23.1

Figure 23.2

Since there is a balance between the number of nuclear positive charges on the protons and the number of equal but oppositely charged electrons orbiting the nucleus, the overall charge on an atom is zero.

Notation, atomic number and atomic mass

General form

$${}^{A}_{Z}X$$

$${}^{12}_{6}C$$ Atomic mass, the number of nucleons (protons + neutrons) in the nucleus

Chemical symbol, in this case **C** for carbon

Atomic number, the number of protons in the nucleus

ATOMIC MASS = NUMBER OF **PROTONS** + NUMBER OF **NEUTRONS**

Or, **A** = Z + **N** where N is the number of neutrons

Worked Example 1

Describe the makeup of the nucleus of the atoms of the following in terms of number of neutrons and protons:

a) ${}^{14}_{7}N$ *b)* ${}^{226}_{88}Ra$ *c)* ${}^{234}_{92}U$ *d)* ${}^{8}_{4}Be$

a) **N**itrogen: **7** protons, **7** neutrons (atomic mass number = **14**)

b) **Ra**dium: **88** protons, **138** neutrons (atomic mass number = **226**)

c) **U**ranium: **92** protons, **142** neutrons (atomic mass number = **234**)

d) **Be**ryllium: **4** protons, **4** neutrons (atomic mass number = **8**)

Ionising radiation

You need to know that the following types of ionising radiation may be emitted from unstable nuclei:

| α | **Alpha particles** are helium nuclei ejected from unstable nuclei. They are heavily ionising and have only a short range, travelling only ~ 10 cm in air. They are stopped by thin card. | ^4_2He |

| β | **Beta particles** are fast moving electrons ejected from unstable nuclei. They are less ionising and travel long distances in air. They are stopped by 1–2 mm of aluminium. | $^{\ 0}_{-1}\text{e}$ |

| γ | **Gamma rays** are photons of high energy EM waves. They are extremely penetrating and interact with atoms which may then emit ionising radiation. They are only stopped by tens of cm of lead. | |

Nuclear transformations

Radioactive forms of some elements (called isotopes) will decay **randomly** over time emitting combinations of the types of radiation mentioned above. The emission of gamma rays has no effect on the atomic mass or charge of the decaying atom (as γ rays are both massless and without charge) but gamma ray emission occurs with other types of radiation.

> Both alpha and beta emissions cause a change in the atomic number of the original decaying element – in both alpha and beta decay processes the original element turns into another element.

Examples of alpha and beta decay

α: An isotope of the radioactive element americium is used in some types of smoke detector; it decays by emitting an alpha particle:

$$^{241}_{95}\text{Am} \qquad \rightarrow \qquad ^{237}_{93}\text{Np} \qquad + \qquad ^4_2\text{He}$$

Note the balanced numbers:

241 = 237 + 4

95 = 93 + 2

An atom of americium-241 decays to an atom of neptunium and emits an alpha particle.

The decay will also involve the production of energy.

β: A radioactive isotope of sodium decays to magnesium by emitting a beta particle:

$$^{26}_{11}\text{Na} \qquad \rightarrow \qquad ^{26}_{12}\text{Mg} \qquad + \qquad ^{\ 0}_{-1}\text{e}$$

Note the balanced numbers:

26 = 26 + 0

11 = 12 − 1

An atom of sodium-26 decays to an atom of magnesium and emits a beta particle.

equations must balance. The sum of the atomic masses before and after the decay process must be the me. The sum of the atomic numbers must be the same before and after the decay.

Note that, for the purpose of balancing nuclear equations, gamma photons have zero mass and zero charge and are sometimes written $_0^0\gamma$. Beta particles have a tiny mass, 0 (which we ignore here), and a charge of -1 and may be represented as $_{-1}^0\mathbf{e}$, as above or $_{-1}^0\beta$.

Beta decay involves a nuclear neutron becoming a proton and an electron which is emitted as the beta particle, thus adding 1 to the atomic number.

Worked Example 2

In the following equation an alpha particle collides with an atom of beryllium transforming it into an atom of carbon-12 and a neutron. Balance the equation.

$$\square\text{Be} \ + \ ^4\text{He} \ \rightarrow \ ^{12}_6\text{C} \ + \ ^1_0\text{n}$$

$$^9_4\text{Be} \ + \ ^4_2\text{He} \ \rightarrow \ ^{12}_6\text{C} \ + \ ^1_0\text{n}$$

Practical work

Measuring the relative penetrating power of the three types of radiation will require the use of a detector like the Geiger-Müller (GM) tube described in the next chapter. Care should be taken when handling even the small samples of radioactive materials used in schools and colleges – open the lead lined storage boxes away from you at arm's length using the handling tongs. These sources should be returned to their storage boxes immediately after use and put back in the lockable storage cupboard.

Chapter 24: Radiation and half-life

Detecting ionising radiation

Nuclear radiation, produced by radioactive isotopes and by the processes in stars, can ionise atoms that it interacts with. This can be detected in a variety of ways. You should know that it can be detected by:

Photographic film: This becomes fogged when exposed to ionising radiation and is used in badges worn by workers at risk from continuous exposure. The badges are checked regularly to ensure that safety limits have not been exceeded.

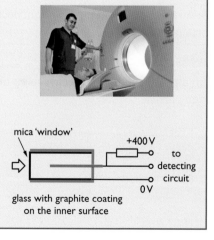

The Geiger-Müller tube: When ionising radiation enters the GM tube it ionises the gases within allowing a pulse of current to pass between the electrodes. This is then fed to either a counter or a rate meter. Often the current pulses are made to produce audible 'clicks'.

mica 'window'

+400 V

to detecting circuit

0 V

glass with graphite coating on the inner surface

Sources of background radiation

You need to remember the following sources of natural or background radiation:

> **From the Earth's rocks**: The slow decay of isotopes of uranium produces radon and thoron gases. Radon is highly radioactive and is a particular problem in certain parts of the UK.
>
> **Cosmic rays**: When stars explode the very violent reactions produce cosmic rays which shower the Earth.
>
> **Medical**: Radioactive materials are used in diagnosis and treatment of illnesses. These contribute to background radiation.
>
> **Nuclear power and weapons**: Testing of nuclear weapons contributes a small amount of background radiation. Leaks from nuclear power stations are also responsible for a small percentage of the background count.

Radioactive decay

Radioactive decay is a random process – it is not possible to predict when an unstable atom will decay or which atoms will decay at any given moment. However, it is possible to predict from previous measurements the percentage of unstable atoms in a sample that will decay in a given time. The rate of decay is measured in **becquerels** – one Bq is one decay per second.

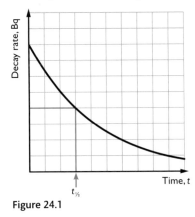

As a decay process proceeds the number of remaining unstable nuclei gets smaller and so too does the rate of decay. The graph, Figure 24.3, shows this. This is an exponential decay – the smaller the number of remaining undecayed nuclei, the more slowly the decay process proceeds.

What does not change is the proportion of the material that decays in a given time.

> The **half-life** of a radioactive isotope, $t_{1/2}$, is the time taken for half the original number of unstable nuclei to decay.

Figure 24.1

The half-life of a particular radioactive isotope does not change. However, different isotopes have half-lives which differ from each other enormously, for example: uranium-238 has a half-life of 4.5 billion years; iodine-131, 8 days; sodium-24, about 15 hours.

Worked Example 3

Iodine-131 has a half-life of 8 days. How much of an initial sample of 400 µg would remain after

a) 16 days, *b)* 24 days, and *c)* 80 days?

The amount will halve every eight days so, 200 µg remain after 8 days,

a) 100 µg after 16 days

b) 50 µg after 24 days

The fraction is $\dfrac{1}{2^n}$ where *n* is the number of half-lives that passed.

c) So after 80 days, or 10 half-lives, the amount remaining would be 0.39 µg

.d be able to measure the background radiation using a GM tube linked to a rate meter. This should
...n into account when measuring the radioactivity of samples of radioactive material. It is important
...member the safety precautions mentioned in the last chapter when handling radioactive materials.

Chapter 25: Applications of radioactivity

Applications of radioactivity

In medicine

Radioactive tracers are used to monitor the function of parts of the body. Chemicals containing gamma
emitters can be swallowed or injected into the body to follow the transport of the tracer through the
digestive system or the veins and arteries. Some compounds are chosen because they are concentrated
in particular organs, allowing the structure and function of the organs to be examined closely.

Radioactive compounds are also used to directly treat illnesses. For example iodine-131, a beta emitter, is
taken orally to treat a condition of the thyroid gland. The iodine is concentrated in the thyroid where the
radiation kills off cells. Focused beams of gamma rays are used to kill off cancerous cells in some tumours.

Sterilisation

Ionising radiation is used to kill off microorganisms and bacteria on surgical equipment and on some
foodstuffs. The items to be sterilised can be sealed in airtight packaging and sterilised through the packaging.

Non-medical tracers

The flow of liquids and gases through industrial processes can be
mapped using radioactive tracers and detectors or gamma cameras.

Please see the Appendix for
additional information on
smoke detectors.

Radioactive dating

By measuring the proportion of the radioactive isotope carbon-14 in samples of dead organic material
it is possible to date how long the material has been dead. Carbon-14 is chemically identical to the
non-radioactive form, carbon-12, so is absorbed by all living things during respiration. The proportion
of carbon-14 to carbon-12 reduces at a predictable rate after death.

Some inorganic materials can be dated by measuring the proportion of a radioactive isotope present relative
to the material formed at the end of its decay chain. Radioactive materials go through series of decays (the
decay chain) transmuting into lighter elements – the end of this sequence is a stable isotope of an element.

Worked Example 4

Gamma emitters are used as medical tracers. Alpha and beta emitters are used to treat tumours. In both cases
isotopes with short half-lives are preferred.

a) Give two reasons why alpha emitters are unsuitable for diagnostic use as tracers.

b) Explain why short half-lives are used in medical diagnosis and treatment.

a) Alpha emitters have an extremely short range inside the body and cannot emerge for detection outside the
body. Alpha particles are extremely ionising and can be very damaging to cells within range.

b) Some of the tracer/treatment materials will be excreted (sweat, urine, etc.) and can therefore pose a health hazard
to others. Prolonged exposure to radiation may cause unwanted damaging effects. A short half-life (a few days or
less) means that the amount of radioactive materials will reduce to an insignificant level within weeks or days.

Hazards of radioactivity

Ionising radiation can kill living cells; if this happens in small numbers and/or to cells that perform non-critical functions, the organism can recover. If the dose is high and particularly if it affects cells which are replaced slowly or perform critical functions, like the nervous system in animals and humans, the damage will be fatal.

Ionisation can affect cells adversely without killing them. If the effect disrupts the genetic material in a cell then the cell may start to malfunction and reproduce uncontrollably resulting in tumours; this is called **cell mutation**. If the cells of critical organs malfunction or if tumours exert excessive pressure on other organs then the effects can also be terminal.

Workers in the nuclear industry are routinely monitored to ensure that their average annual exposure is within safe limits. Badges which discolour with exposure to radiation give an immediate warning of possible overexposure. Radiation sources are handled either with thick gauntlets or with remote controlled handling devices for very radioactive materials. People who live in areas where the exposure to natural radiation from rocks is high should ensure their houses are ventilated to ensure the build up of the radioactive gas radon does not occur.

Fission reactors used in the production of electricity produce a variety of radioactive waste products. Some have short half-lives and low activity. Other waste materials are highly radioactive and have long half-lives. Such materials need to be encapsulated and stored in places far away from human contact. Safely enclosing these waste products, that are gamma emitters, requires lead, concrete or both to ensure no radiation escapes. The long half-lives of some products mean that the waste needs to be stored securely (no leakage, no theft) for tens or even hundreds of years.

Safe handling of radioactive materials

Safe handling procedures have been mentioned at the end of Chapter 23. Radioactive materials must be clearly marked as shown in Figure 25.1 and stored securely.

Handling highly radioactive materials is done remotely using electronically controlled grippers viewed through thick lead glass or viewed with suitable protected cctv equipment.

Figure 25.1

Practical work

None relevant to this chapter.

Chapter 26: Particles

The Rutherford model of the atom

Geiger and Marsden's experiment

This is a key experiment in the development of Rutherford's model of the atom.

A beam of alpha particles was fired at a gold leaf (see Figure 26.1, overleaf). A zinc sulphide screen was used as a detector. Key observations were:

> Most of the alpha particles passed through without recoiling (bouncing back) (A) or deflecting off gold atoms. Some deflected as they passed through the gold leaf (B) and a tiny proportion bounced back (C).

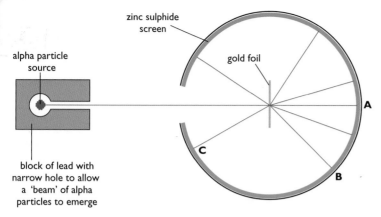

Figure 26.1 *Firing alpha particles at thin gold foil*

The deflections were caused by the electric repulsion between the positive charges of the alpha particles and the positive charges in the gold nucleus. The amount of the deflection depended on a number of factors:

- The speed of the alpha particles – faster particles are deflected through smaller angles.
- The nuclear charge – more highly charged nuclei (like gold with 79 protons) produce greater deflections.
- How close the alpha particle gets to the nucleus – the electric force diminishes with the square of the distance.

Rutherford's deductions

The atom is mostly empty space – the positive charge must be concentrated in a very small central region of the atom. (Previously it was thought that the positive charge was uniformly distributed throughout the atom.) From the observations of the scattering experiment Rutherford deduced that the nucleus was one ten-thousandth of the diameter of the atom.

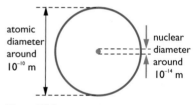

Figure 26.2

Nuclear fission

The isotope uranium-235 is **fissile** – this means it can be split into lighter elements quite easily. If an atom of U-235 is struck by a low energy neutron (relatively low speed) it breaks apart as shown in Figure 26.3 releasing energy in the form of the kinetic energy of the decay fragments (and gamma ray photons).

The **parent** nucleus, U-235, produces two **daughter** nuclei as shown plus a number of neutrons. If these neutrons are absorbed by further fissile nuclei then a **chain reaction** results with rapidly increasing numbers of atoms splitting apart and releasing energy – a nuclear bomb is an uncontrolled chain reaction.

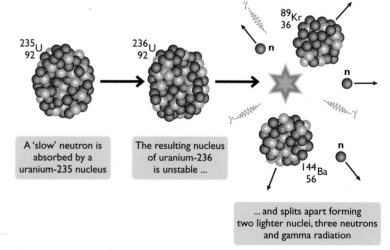

A 'slow' neutron is absorbed by a uranium-235 nucleus

The resulting nucleus of uranium-236 is unstable ...

... and splits apart forming two lighter nuclei, three neutrons and gamma radiation

Figure 26.3 *Fission of uranium-235 triggered by a slow neutron.*

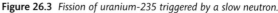

Practical work

None relevant to this chapter.

Generating electricity with nuclear fission

If the chain reaction in a fissile nuclear fuel like uranium-235 is controlled, so that the energy is released much more slowly, the heat produced can be used to generate steam to turn turbines and drive generators, as shown in Figure 26.4.

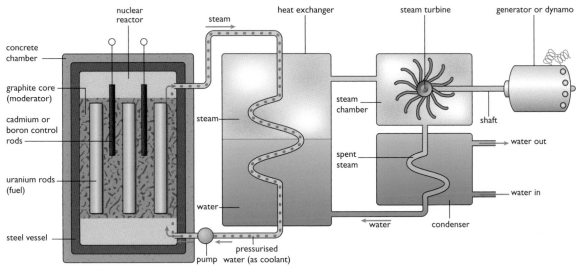

Figure 26.4 *A schematic diagram of a nuclear power station*

Key points:

The graphite **moderator** is used to absorb some of the energy of fast neutrons so they are readily absorbed by nuclei of U-235 sustaining the chain reaction.

The boron **control rods** absorb neutrons to take them out of action completely. The control rods can be raised out of the reactor core allowing the chain reaction to speed up, or lowered completely to shut down the chain reaction.

Warm-up questions

1 Copy and complete the table comparing the properties of the three principal particles that make up an atom.

Particle	Relative Mass	Relative Charge	Location
		+1	In the nucleus
Neutron	1		
	$\sim \dfrac{1}{2000}$		

2 Define the following terms:
   a) atomic number   b) atomic mass   c) isotope

3 Uranium-238 and Uranium-235 are two isotopes of uranium. How is the composition of an atom of each isotope: a) the same and b) different?

4 Complete the following table to describe the structure of different isotopes of the elements:

Isotope	Element	Protons	Neutrons	Electrons
$^{239}_{92}U$	Uranium			
$^{14}_{6}C$	Carbon			
$^{\square}_{56}Ba$	Barium		83	

5 When an unstable atom decays it may emit a number of different types of radiation; α, β and γ. Explain what each of these is.

6 α, β and γ are all types of *ionising* radiation.

   a) Explain what is meant by ionising radiation.

   b) Compare the ionising power of each kind of radiation.

7 Describe an experiment to determine what kind(s) of nuclear radiation a radioactive isotope was producing. List the apparatus needed and describe the experimental set-up. Explain how you would interpret the results of your investigation.

8 Copy and balance the following nuclear equations:

   a) $^{235}_{92}U + ^{1}_{0}n \rightarrow 3\ n + ^{139}Ba + _{36}Kr$
     [This could be the start of a chain reaction.]

   b) $^{1}H + ^{3}_{1}H \rightarrow\ ^{\ }He$
     [This is the fusion process that occurs in the Sun.]

   c) $^{27}_{13}Al + ^{1}_{0}n \rightarrow\ ^{24}Na + \ _{2}$
     [This is an alpha decay process.]

   d) $^{239}_{92}U + ^{4}He \rightarrow\ _{94}Pu + \ n$

9 State two methods used to detect ionising radiation.

10 State two main sources of *background* radiation in the UK.

11 a) Radioactive decay is measured in becquerels; define the becquerel.

   b) Sometimes the rate at which water runs out of a container or burette is used to model radioactive decay. What particular aspect of radioactive decay does the model show?

12 a) Different radioactive isotopes have different *half-lives*. Explain what is meant by the half-life of an isotope.

   b) Isotope A has a half-life of 2 days and isotope B has a half-life of 10 days. The initial amount of both these isotopes is 320 µg. How much of each isotope remains after 10 days?

   c) Is it possible to determine which particular atom of an isotope will decay if you know the half-life of the isotope? Give a reason for your answer.

13 Iodine-131 is a radioactive beta emitter with a half-life of 8 days. Like the common stable form of iodine, it is concentrated in the thyroid gland and excess iodine is excreted in urine and in sweat.

   a) Why is iodine used in the treatment of cancer of the thyroid?

   b) Why is it particularly suited for treating thyroid cancer?

   c) Why are patients, who are undergoing radio iodine treatment, told to avoid sharing both toilet and washing facilities with other people?

14 List three dangers associated with the use and handling of radioactive materials producing ionising radiation.

15 a) Copy and complete the diagram below which shows the uncompleted paths of alpha particles fired at gold foil in Geiger and Marsden's experiment.

   b) Explain why the alpha particles follow the paths you have shown.

   c) How would your answer have differed if the alpha particles had less KE?

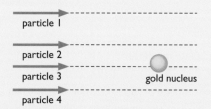

16 What deductions did Rutherford make about the nucleus of a gold atom from the findings of the experiment carried out by Geiger and Marsden, mentioned in question 14?

17 Uranium-235 is an example of a *fissile* material. Explain what this means.

18 Copy and complete the following sentences about the fission of uranium-235:

When an atom of uranium-235 splits two smaller _____ are produced together with two or three _____ and some _____. The original nucleus that splits is called the parent and the two fragments are called the _____ _____.

19 Uranium-235 fission is triggered by the impact of a particle. Name this particle and describe its speed. Describe how a chain reaction may be triggered in a fissile material like uranium-235.

20 Nuclear power stations use a controlled chain reaction to generate heat to boil water and ultimately drive a turbine. The reactor core is encased in a steel vessel surrounded by concrete and contains a *moderator* into which the *fuel rods* are lowered. Another set of rods called *control rods* can be lowered into the core or raised out of it.

a) Why is the reactor vessel surrounded by thick concrete?

b) Explain the purpose of the moderator.

c) What do the fuel rods consist of?

d) What is the purpose of the control rods?

e) In the event of the reactor core starting to overheat, what action should be taken to shut down the chain reaction?

Notes

Exam-Style Questions

Section A: Forces and Motion

1 A marathon runner covers the *distance* of 42 km in a *time* of 3 hours and 20 minutes.

  a) State, in words, the formula you should use to calculate average speed of the runner. *[1]*

  b) Use your formula to calculate the average speed of the runner,

    (i) in km/h, and **(ii)** in m/s. *[3]*

  c) The graph below shows how the runner's speed changed during the race:

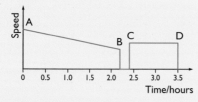

  Describe the way the runner is running during the period from

    (i) A to B **(ii)** B to C and **(iii)** C to D. *[3]*

  d) Explain how you would use the graph to calculate the distance the runner ran in the period AB. State the additional information you would need. *[2]*

2 a) Velocity is a *vector* quantity; speed is a *scalar* quantity. Explain the difference between a vector and a scalar quantity. *[3]*

  b) Give one example of:

    (i) a scalar quantity, and

    (ii) a vector quantity. *[2]*

  c) A student calculates the acceleration of a car and finds that it is 1.8 m/s². Explain what this means. *[2]*

3 The diagram below shows a tightrope walker standing on a tightrope.

  a) Label the forces that are acting on the tightrope walker with three clearly labelled arrows. *[5]*

  b) The tightrope walker is stationary. Explain what this tells you about the forces acting on the tightrope walker. *[2]*

  c) Why is it impossible for the tightrope to be perfectly horizontal when the tightrope walker is standing on it as shown in the diagram? *[2]*

4 Sharlini measures the length of a light spring and records this length I_o. She then adds increasing amounts of known mass onto the spring and measures the new length of the spring for each value of the load. Her results are shown in the table below.

Natural length, I_o	3.2 cm

Load in g	Length of spring in cm
100	4.8
200	5.4
300	5.9
400	6.6
500	7.2
600	7.9
700	8.7

Sharlini wants to plot a graph to show that the spring obeys Hooke's Law.

  a) Explain how she should use her results to do this. *[4]*

  b) Plot a suitable graph to present her results. *[4]*

  c) Do her results show that the spring obeys Hooke's Law? *[3]*

  d) At the end of her experiment, Sharlini notices that the spring now has a natural length of 3.4 cm. Comment on this. *[2]*

5 A toy car rolls down a gently sloping ramp. It has a tape attached to it which it pulls through a ticker-timer as it rolls down the slope. The ticker timer is designed to make a mark on the tape every tenth of a second. A length of the tape is shown below.

  a) What is the average speed of the car? *[2]*

  b) Describe the motion of the car. *[1]*

The slope of the ramp is changed twice and two further lengths of tape are obtained.

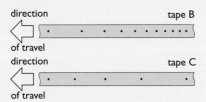

c) (i) What does tape B reveal about the motion of the car?

(ii) What has been done to the ramp to cause the change in the motion of the car? **[3]**

d) (i) What does tape C reveal about the motion of the car?

(ii) What has been done to the ramp to cause the change in the motion of the car? **[3]**

In one further experiment measurements made from the tape show that the car, which has a mass of 1200 g, is accelerating at 4.5 cm/s².

e) Calculate the resultant force on the car, showing your working. **[3]**

6 This question is about a car driver making an emergency stop. Look at the speed–time graphs shown below.

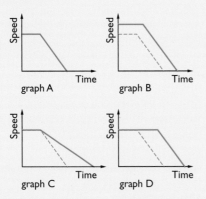

graph A   graph B

graph C   graph D

Graph A shows the speed–time graph for an alert driver driving at the legal speed limit, and braking without skidding, in good conditions. (Graph A is superimposed on the other 3 graphs in green for comparison.)

a) This graph, and the others that follow, can be divided into two sections. State what these two sections are and identify them on the graph A. **[4]**

b) Suggest why it takes longer to stop in the other three cases. **[6]**

c) Graphs B, C and D reveal that it will take longer for the car to come to rest and it will have travelled further while doing so. Explain how you could work out the total stopping distance if scales were marked on both axes of graph B. **[3]**

7 A Saturn V rocket (the type used during the manned missions to the Moon) had a mass of 3 million kilograms. The initial thrust provided by the rocket motor was 34 000 000 N.

a) Calculate the acceleration of the rocket on take off. **[3]**

b) The velocity needed to break free from the pull of the Earth's gravity is 11.2 km/s; assuming the acceleration you have just calculated remained constant, how long would it have taken for the rocket to reach this 'escape' velocity? **[3]**

c) Give one reason why the acceleration of the rocket did not remain constant. **[2]**

8 The graph below shows the velocity–time graph for a free fall parachutist. Describe and explain the features of the graph. In your explanation you should consider the forces that act on the parachutist.

Note that there are 3 distinct phases to the motion, indicated by the times t_1, t_2, and t_3. **[9]**

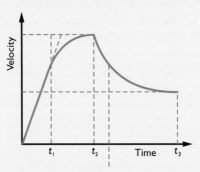

9 a) Write a word definition of momentum. **[1]**

b) What is meant by the law of conservation of momentum? **[2]**

c) A supermarket trolley of mass 20 kg travelling at 5 m/s collides with some stationary trolleys that have been stacked together. The moving trolley joins up with the others and the group moves at 1.25 m/s after the collision. Assuming that the trolleys are identical and we can ignore external forces like friction, how many trolleys were there in the stationary stack before the collision? **[4]**

10 a) Explain the difference between the following types of collision: elastic, inelastic, partially elastic. *[4]*

b) Give an example of **(i)** an inelastic collision, and **(ii)** a partially elastic collision. *[2]*

11 The diagrams below show a collision between two balls on a smooth surface. (Smooth means that friction between the balls and the surface can be ignored.)

red ball
at rest
$U_R=0$

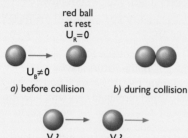

$U_B\neq 0$

a) before collision   *b) during collision*

$V_B?$ $V_R?$

c) after collision

a) Newton stated three laws about the way bodies react to forces:

1: Objects keep moving in a straight line or remain at rest unless acted on by a resultant force.

2: The rate of change of momentum of a body is proportional to the force that acts on it.

3: For every action there is an equal and opposite reaction.

(i) Explain in which diagram or diagrams law 3 applies.

(ii) Explain in which diagram or diagrams law 1 applies.

(iii) Use law 2 to write equations for change in momentum of each ball as a result of the collision.

What relationship is there between the change in momentum of each ball? *[7]*

b) The balls have the same mass and the velocity of the red ball after the collision, V_R, is the same as the velocity of the blue ball before the collision, U_B.

(i) What is the velocity of the blue ball after the collision, V_B? Explain your answer.

(ii) Is the collision inelastic, partially elastic or elastic? Explain your answer. *[4]*

12 Explain, in terms of momentum change, how each of these safety features can help to reduce the risk of injuries in a car crash.

- Seat belt
- Air bag
- Crumple zone *[4]*

13 This question is about Newton's 1st Law of Motion. In each of the following situations state whether the object is subject to a resultant force and, if it is, state what is responsible for the resultant force.

a) A communications satellite in orbit around the Earth.

b) A trampolinist at **(i)** the bottom of a bounce, and **(ii)** the top of his bounce.

c) A car travelling at a steady speed along a straight motorway. *[6]*

14 A ballistic balance is used to measure the speed of a bullet. The set-up for this apparatus is shown below.

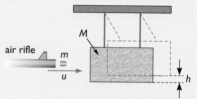

air rifle M m u h

An air rifle fires a pellet of mass **m** grams, horizontally with velocity **u** m/s into a block of absorbant material of mass **M** kg. This mass is freely suspended. Momentum is conserved during the collision and the block with the embedded pellet moves off horizontally with a horizontal velocity of **v** m/s immediately after the collision. The block swings through an arc and comes to rest for an instant at a height of **h** metres above its original position.

a) You have a balance which can weigh masses up to 1 kg to an accuracy of 1 g. The pellet has a mass less than 1 g. Explain how you would measure the mass of a pellet accurately. *[2]*

b) What type of collision takes place when the pellet is fired into the block? *[1]*

c) If no energy is lost in the impact how can the velocity, **v**, be calculated from the measurement made on the block's movement after the impact? *[3]*

d) Given you have calculated a value for **v**, and that you have measured **m** and **M**, explain with the aid of a formula how you would calculate the velocity, **u**, of the pellet. *[3]*

e) The assumption that all the energy of the pellet is transferred to the block is not strictly accurate. Explain why and state whether the value calculated for **u** will be too high or too low as a result. *[3]*

15 Two ice skaters, Jane and Chris, stand facing one another. Jane has a mass of 50 kg and Chris has a mass of 75 kg. Jane pushes Chris with a horizontal force of 10 N for a time of 0.5 s.

a) Describe what you would expect to see happen to Jane and Chris after this has happened? [3]

b) Explain, with reference to Newton's Laws, why what you would see happening occurs. [3]

16 a) Explain, with the aid of a clear labelled diagram, how you would calculate the moment of a force. [3]

b) The force on a concrete block is upwards as shown. What can you say about the force the block exerts on the short end of the crowbar? [2]

c) Calculate the size of the upward force (the LOAD) on the concrete block shown in the diagram. An effort force of 30 N is applied acting vertically downwards on the crowbar as shown. [3]

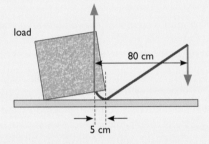

load

80 cm

5 cm

17 a) Double decker buses are subjected to a 'topple test' to see how far they they can tilt before they fall over. They are tested in the worst possible condition for stability with a full load of passengers (dummies of the same mass) on the upper deck and with no passengers on the bottom deck. How does this loading affect the stability of the bus and why is the test carried out in this way? [3]

b) You are designing the base for a large sun umbrella to be used to provide shade for people sitting around a hotel swimming pool. Explain how you would make it stable enough not to topple over on a windy day. [3]

18 A wooden bench consists of a plank which weighs 80 N resting on two brick pillars. The bench is 2.5 m long and the pillars which support it are 0.5 m from each end, as shown.

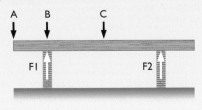

A   B   C

F1   F2

a) Mark an X on the diagram where you would expect the centre of gravity of the plank to be. [1]

b) The brick pillars provide upward forces F1 and F2 on the bench. What will the sizes of these two forces be when **(i)** no one is sitting on the bench, and **(ii)** a child who weighs 200 N is sitting in the middle of the bench in position C. [4]

c) Describe how the two forces, F1 and F2, will change if the child moves from position C to position B and then to position A. [5]

19 Identify the following objects found in the universe from their descriptions:

a) Goes round the Sun with a very elongated (eccentric) orbit and has a distinctive tail.

b) One or more of these may orbit a planet.

c) This consists of the Sun and the planets.

d) The Milky Way is an example of one of these; billions of them make up the universe.

e) These provide huge amounts of energy burning nuclear fuel in fusion reactions. [5]

20 Copy and complete this sentence:

Objects with _____ exert a _____ force on each other. It is this force that keeps the planets in orbit around the Sun. This size of this force depends on the _____ of each object and the _____ they are apart. This force gets _____ as the _____ between the two objects _____. [7]

21 The Moon travels around the Earth in a nearly circular orbit of average diameter 385 000 km once every 27.4 days. Calculate the orbital speed of the Moon in m/s using the formula given at the front of the exam booklet. [4]

22 A communications satellite mantains its position above the equator by orbiting the Earth at the same rate that Earth rotates. This is called a geostationary orbit. To do this the orbital speed of the satellite must be 3.1 km/s. The Earth has a radius of 6400 km. Use this information to calculate:

a) The radius of the orbit of the geostationary communications satellite.

b) How high the satellite is above the surface of the Earth. [4]

Section B: Electricity

1 Here is a diagram of an electric plug which has been wired incorrectly:

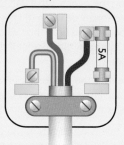

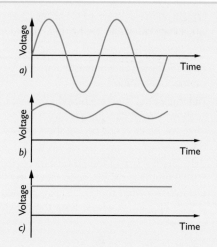

   a) State the error that has been made in wiring the plug. **[1]**

   b) Will the appliance to which this plug is connected cause a fuse to 'blow' when it is plugged into the mains supply? **[1]**

   c) Will it cause a RCD (power breaker) safety device to trip when it is plugged in? **[1]**

   d) What is the purpose of the fuse included in the plug? **[2]**

   e) This plug is connected to a 2 kW electric fire designed to run from the 230 V AC mains supply. Calculate normal current that the fire will draw, showing your working. Is the plug correctly fused? **[3]**

2 The energy consumed by electrical appliances is usually measured in units called kilowatt-hours, kWh. This is the energy used by a 1 kW appliance in one hour.

   a) Express the energy unit kWh in terms of the standard unit of energy, the joule. **[2]**

   b) A tumble drier with a heater of 1.2 kW is used twice a week for 90 minutes on each occasion. How many kWh of energy are consumed by the tumble drier in a year? **[4]**

   c) A reading lamp draws a current of 255 mA when operated from a 230 V AC mains supply. What is the power rating of the lamp? How long would it take in hours to consume one unit (1 kWh) of electrical energy? **[4]**

3 The following graphs show how voltage varies with time for 3 different sources of electrical energy (the graphs do not necessarily have the same scales):

The graphs represent the voltage–time graph for a battery, for an AC supply and for the output of a microphone. State which graph is which. **[2]**

4 *a)* Explain how insulators like plastic may be charged with static electricity. **[3]**

   b) Describe how you would demonstrate by experiments that different types of plastic charge up with different types of static electricity. **[4]**

   c) You cannot charge a metal ruler in the same way as a plastic strip; explain why this is so. **[2]**

5 *a)* Give two examples of static electricity in use in everyday situations. **[2]**

   b) Give two examples of hazards caused by the build-up of static charge on objects. **[2]**

6 *a)* What physical feature makes certain materials very good electrical conductors and others non-conductors? **[2]**

   b) State the units of electric current and electric charge. **[2]**

   c) State the relationship between current and charge. **[2]**

   d) A filament lamp in a torch draws a current of 50 mA. If the torch is switched on for half an hour, how much electric charge will be circulated by the battery? **[3]**

7 *a)* When there is a voltage difference across a resistor a current will pass through it. As the current flows through the resistor electrical energy is transferred to the resistor in the form of heat. Define the unit of voltage in terms of energy transfer to the resistor. **[2]**

b) **(i)** If a current of 0.5 A flows through a resistor for 100 s how much charge has passed through the resistor?

(ii) If the voltage across the resistor is 12 V how much energy has been transferred to the resistor? **[4]**

8 State what you expect the ammeters and voltmeters to read in each of the following circuits:

(The readings on some of the meters are given, only give answers for the unlabelled meters.) **[7]**

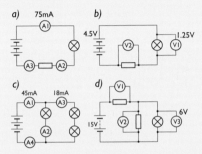

a) 75mA   *b)*

c) 45mA   18mA   *d)*

9 *a)* State, in words, the relationship between the current through a resistor and the voltage across the resistor. **[1]**

b) Look at the circuit shown above right. State the name of component X. **[1]**

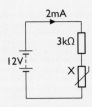

c) Calculate the voltage across the 3 kΩ resistor. **[2]**

d) State the voltage across component X. **[1]**

e) Calculate the resistance of component X. **[2]**

f) A hairdrier blowing hot air is held close to component X. What effect will this have on component X and how will this affect **(i)** the current in the circuit, and **(ii)** the voltage across component X? **[4]**

10 *a)* **(i)** What is an LED?

(ii) Draw the circuit symbol for an LED. **[2]**

b) A 'helpful' pupil decides to test LEDs for a teacher using a 6 V battery and some connecting wire. He reports back that none of the LEDs he has tested works; some lit once briefly and brightly, then went out, others did not light at all. Given that the LEDs were brand new, explain the results of the pupil's test and explain how the test ought to have been carried out safely. **[6]**

Section C: Waves

1 A set of waves on the water in a ripple tank is made by vibrating a bar in contact with the water surface. These waves spread out across the water surface. Here are two students' results; the diagrams show a cross-section through the water surface at an instant in time. Assume the waves travel at the same speed in both students' experiments.

a)

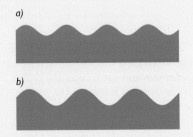

b)

a) **(i)** Mark the *wavelength* clearly on each diagram.

(ii) Explain how waves with a different wavelength are produced. **[4]**

b) **(i)** Mark the *amplitude* of the waves clearly on each diagram.

(ii) Explain how waves with a different amplitude are produced. **[4]**

c) **(i)** State whether these waves are *transverse* or *longitudinal*. **[1]**

(ii) Give an example of another wave of this type. **[1]**

d) **(i)** Describe a method you would use to *measure* the frequency of the waves in an experiment like this. Describe any measuring equipment you would use and explain how you would use your measurements to find the frequency. **[4]**

(ii) In an experiment like this the frequency of the waves is measured and found to be 7 Hz. The wavelength is measured and found to be 4 cm. Calculate the speed the waves are travelling across the ripple tank. Show any formula you use and your working. **[3]**

2 a) Explain with the aid of a labelled diagram what is meant by the *diffraction* of waves. **[3]**

b) The student sets up a ripple tank demonstration of diffraction using the apparatus shown below.

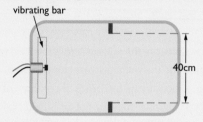

vibrating bar

40cm

The period of the vibrations of the bar is controlled by the current supplied to the motor. In the student's experiment the period is 0.02 s and the wavelength of the ripples is 0.5 cm. When the experiment is carried out the results do not show the expected effect.

Give details of *two* things that the student could change in his experiment to make the effect much easier to see. **[4]**

3 a) Sound waves and light waves are different in a number of ways. Avoiding the obvious (you cannot hear light waves, etc.) state *two* ways in which they are different. **[2]**

b) In a simple experiment to measure the speed of sound a short sharp sound is made at a distance from a large reflective surface like a wall of a large building, to produce a good echo. Describe what measurements you would take to calculate the speed of sound by this method. You should state the measuring instruments you would use and describe how you would try to make the measurements as accurate as possible. You should also explain how you would calculate the speed of sound using your measurements. **[6]**

c) Why is it difficult to measure the speed of light in a similar way, using a mirror, in a school physics lab? **[2]**

4 Complete the following diagram which shows the visible spectrum, partially labelled.

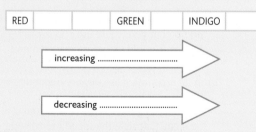

| RED | | | GREEN | | INDIGO | |

increasing ....

decreasing ....

5 Describe, with the aid of clearly labelled diagrams, *analogue* and *digital* signals. **[4]**

6 a) Here are four diagrams showing how a ray of light is reflected from a flat mirror surface. State which one(s) is(are) correct. **[1]**

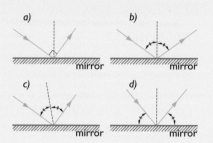

a) mirror   b) mirror
c) mirror   d) mirror

b) A ray of light strikes a mirror as shown in A. The direction of the ray of light remains the same but the mirror is turned through an angle of 10° as shown in B.

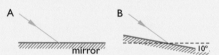

A   mirror   B   10°

(i) What is the *increase* in the angle of incidence that results from this? **[1]**

(ii) If the angle of incidence increases by this amount, what happens to the angle of reflection as a result? **[1]**

(iii) What is the total angle that the reflected ray is turned through as a result of turning the mirror through an angle of 10°? **[2]**

7 Complete the following diagram to show how a ray of light passes through a rectangular glass block. Draw and label the normal. Label the incident ray and the refracted ray. Label the angle of incidence *i*, and the angle of refraction, *r*. Show how the ray emerges from the block. **[7]**

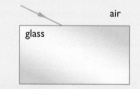

air
glass

8 a) Draw a ray of light meeting the boundary between glass and air so that it is *just* totally internally reflected. Label the critical angle on your diagram and show the path that the reflected ray takes. **[4]**

b) (i) The refractive index for a particular type of glass (in air) is 2. Write a formula to show how refractive index is related to the critical angle. **[2]**

(ii) Use your formula to calculate the critical angle for this type of glass. **[2]**

9 **a)** State the range of of frequencies of sound audible to humans. **[2]**

b) The following diagram shows a sound wave that has been converted to an electrical signal and displayed on an oscilloscope.

The voltage per division is: 20 mV/div. The time per division is: 5 ms/div

(i) State the amplitude of the signal waveform. **[2]**

(ii) State the period of the signal waveform. **[2]**

(iii) Calculate the frequency of the sound that produces this waveform on the oscilloscope. **[2]**

c) Sketch what you would expect to see on the oscilloscope screen if the sound was made quieter and lower in pitch. **[2]**

Section D: Energy Resources and Energy Transfer

1 **a)** State the law of conservation of energy. **[2]**

b) 1 litre of petrol will produce approximately 30 million joules of energy when burnt.

(i) If it is used to power a small electric generator which is 60% efficient how much useful energy is produced? **[2]**

(ii) State two unwanted energy conversions which take place. **[2]**

(iii) Draw a Sankey diagram to represent this energy conversion process. **[3]**

c) The electrical output of the generator is used to run a refrigerator that has a power rating of 250 W. Calculate how long it can run on the generator output from 1 litre of petrol. **[3]**

2 **a)** Complete the following sentences about heat transfer:

Thermal _____ is the transfer of _____ _____ through a substance without the substance itself _____. Heat energy is transferred in fluids by _____; as the fluid is heated it expands and become less _____ and the warmer fluid is displaced _____ by colder _____ fluid. Thermal _____ is the transfer of energy in the form of _____ _____; this heat transfer mechanism does not require a _____ medium. **[12]**

b) In an experiment to test the best way of keeping a drink warm, Liam devises three sets of apparatus, shown below:

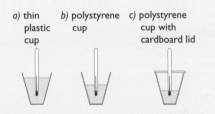

a) thin plastic cup   b) polystyrene cup   c) polystyrene cup with cardboard lid

He timed how long it took for the contents of each cup to cool down by the same amount.

He had hoped to demostrate that one particular improvement on the basic thin plastic cup would have the best effect on keeping the tea warm, but his results were inconclusive.

Explain what he did wrong with his experiment and describe a fair way of testing the effect of different types of insulation. **[6]**

3 A weight lifter lifts a barbell of mass of 50 kg from chest height to above his head (a distance of 75 cm) 15 times in 1 minute.

a) Calculate the weight of the barbell. **[1]**

b) Calculate the work done in lifting the barbell once. **[3]**

c) Calculate the power developed by the weightlifter in this set of 15 lifts. **[3]**

4 A hydroelectric power station uses surplus energy produced at certain periods of the day to pump river water from a pumping station in the valley back to a reservoir in the mountains. The reservoir is 600 m above the pumping station. During one pumping session 10 000 litres of water are pumped up to the reservoir. [Take the mass of 1 litre of water as 1 kg.]

a) State the energy conversions that take place in this process. **[4]**

b) Calculate how much energy is given to the water in the pumping session. **[3]**

c) Explain why the process of moving water in this way cannot be 100% efficient. (Hint: think about the energy conversions.) **[3]**

d) Given that both the processes of pumping the water up to the mountain and then, at some later time, converting the energy back into electricity are less than 100%, explain why is this done. **[2]**

5 Electricity can be generated in different ways. Name an example of:

a) A fossil fuel that is used to generate electricity.

b) A renewable energy source that is used to generate electricity.

c) An energy source that cannot deal with a steady demand for electricity.

d) An energy source that can only be utilised in cetain places. **[4]**

6 Scotland is an ideal place for the generation of hydroelectricity.

a) Give two reasons why this is so. **[2]**

b) Scotland is a long way from major industrial centres in England. Give a reason why this is a drawback for the generation of hydroelectricity. **[2]**

c) Explain why the first nuclear power plant was constucted in the far north of Scotland. **[2]**

Section E: Solids, Liquids and Gases

1 a) Write a word equation for calculating the density of a substance. **[2]**

b) Martha does an experiment to measure the density of liquid. In this experiment she measures the volume of the liquid in a measuring cylinder calibrated in cm³ and then she weighs the measuring cylinder on an electronic balance calibrated in grams.

She does this for eight different amounts of liquid.

(i) The figure shows an example of the measuring cylinder she used. State the volume of liquid in the cylinder. **[2]**

(ii) Here are her results.

Use her results to plot a graph of mass against volume. **[6]**

(iii) One of her results looks a little strange. State which one and give a possible reason for it. **[2]**

(iv) She expected the graph to pass through the 0,0 point on her scale, but it does not. State why it ought to pass through this point and give a likely reason why it did not. **[2]**

(v) Use Martha's graph to calculate the density of the liquid she used in this experiment. **[4]**

(vi) State why it is better to produce a graph of a range of results rather than just use one set of mass and volume readings. **[2]**

2 A reservoir is made by building a dam across a river. Here are some possible cross-sections of shapes that might be used for the dam:

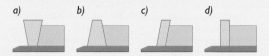

State which shape is the most suitable and give your reason. **[3]**

3 The figure opposite shows a 5 metre cube completely submerged in water, at a depth of 10 m below the water surface.

Volume of liquid	10	20	30	40	50	60	70	80
Mass of liquid	42.1	54.0	65.9	78.5	90.0	102.0	113.9	126.1

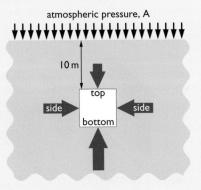

atmospheric pressure, A

10 m

top
side   side
bottom

Note: The cube is **not** drawn to scale

a) Water exerts a pressure on both sides of the cube as shown. Explain why there is no resultant horizontal force on the cube. [3]

b) The density of water is 1000 kg/m³. Calculate the extra pressure (in addition to the pressure of the atmosphere) that acts on:

(i) the top surface of the cube, and

(ii) the bottom surface of the cube. [4]

c) Use your answers to b) to calculate the total force acting:

(i) downwards on the top surface, and

(ii) upwards on the bottom surface. [4]

d) State the size and magnitude of the force which acts on the cube as a result of the pressure of the water. State the name of this force. [3]

e) Can you say whether the cube will move up or down or remain stationary? Give your reasons. [3]

4 Complete the following sentences about the effect of heat on ice taken from a freezer:

Initially the temperature of the ice will _____.
When it reaches the _____ point of 0°C the ice will begin to _____. When all the ice has changed from the _____ state to the _____ state, its temperature will continue to _____ as a result of being heated. During this time some of the _____ will turn to a _____; this process is called _____.
When the _____ point of 100°C is reached the temperature will stop rising until all the_____ has turned to _____. [12]

5 a) Label this diagram of a simple experiment to demonstrate Brownian motion in air. [3]

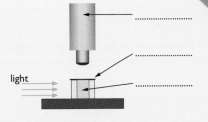

light

b) Describe, with the aid of a diagram, what you would expect to observe. [3]

c) This experiment led to a better understanding of the way gases behave. State the conclusions about the motion of air molecules that were drawn from the observations of Brownian motion. [3]

6 In an experiment to investigate the relationship between the pressure and volume of a fixed amount of gas at constant temperature, a class produced the following graphs of pressure against volume:

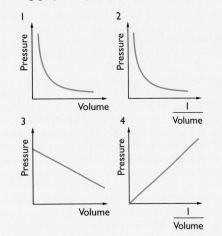

Which answer below lists the graphs that accurately represent the relationship?

A: none of them

B: 1 and 3

C: 2 and 3

D: 1 and 4

E: 1 only [2]

7 a) The air in a rigid steel container is at a pressure of 50 kPa when the temperature is – 23°C. The container has a safety valve which releases the gas when the pressure in the container reaches 300 kPa. When heated, at what temperature, in °C, will the pressure valve release? [4]

b) Explain why, in terms of the behaviour of the air molecules, the gas exerts a different pressure on the walls of the container as the temperature is changed. [2]

Section F: Magnetism and Electromagnetism

1   State which of the following field patterns is correctly drawn and explain what is wrong with the others.

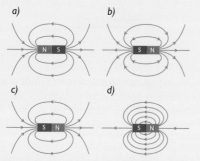

2   **a)** What type of magnetic material is most suitable for the core in an electromagnet, steel or 'soft' iron? Give a reason for your answer. **[2]**

   b) Describe the structure of an electromagnet with a labelled diagram. **[3]**

   c) State two ways in which the electromagnet could be made more powerful. **[2]**

3   Fleming's Left-Hand Rule is used to work out how a wire with a current passing through it reacts when it is placed in a magnetic field.

   a) How are the first two fingers and thumb of your left hand arranged when you apply this rule? **[2]**

   b) Label the figure to show which part of the hand represents which quantity when you are trying to predict how the current-carrying wire will behave in a magnetic field. **[3]**

   c) Use Fleming's LH Rule to predict how passing a current through the coil affects the coil shown above, right, and label the diagram clearly to explain what happens. **[4]**

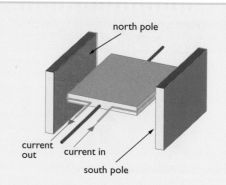

north pole

current out   current in

south pole

4   **a)** In which of the following situations is a current induced in the coil attached to the sensitive ammeter? Explain your answer. **[4]**

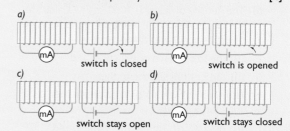

   b) A bicycle dynamo consists of a magnet that is made to rotate inside a coil of wire.

   (i) What energy conversion does the dynamo perform? **[2]**

   (ii) State three ways in which the voltage induced by a dynamo can be increased. **[3]**

5   **a)** Label the following simple diagram of a transformer. **[3]**

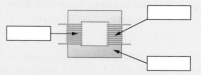

   b) What is the purpose of a transformer and what feature of its design achieves this? **[4]**

   c) Transformers are assumed to be 100% efficient. Explain what this means in terms of electrical input power and electrical output power with a simple equation. Define the terms in your equation clearly. **[4]**

   d) Transformers are used in the generation and distribution of electricity to reduce energy loss caused by the resistance of the transmission lines. Explain, in simple terms, how this is achieved. Your answer should refer to your answer to part *c)*, above. **[4]**

Section G: Radioactivity and Particles

1 **a)** Label the following simplified diagram of an atom, to show the main particles of which it is formed. **[3]**

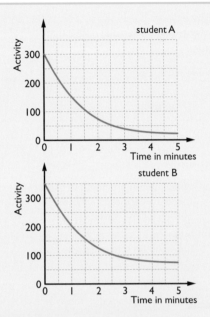

b) (i) State which particles make up the nucleus. **[2]**

   (ii) State the collective term for particles found in the nucleus of an atom. **[1]**

c) The notation used in nuclear equations is shown below, with X representing the symbol for the element in question. Define the terms A and Z.

$$^A_Z X$$ **[2]**

2 **a)** Radioactive decay is a random process. Explain what this means. **[2]**

b) When an unstable atom decays it may emit different types of *ionising* radiation. State the meaning of ionising. **[2]**

c) State the names of the three principal types of ionising radiation that may be emitted by an unstable atom when it decays. **[3]**

d) Identify the types of radiation referred to in the following descriptions:

   (i) 'This type of radiation consists of a high speed electron that is emitted from the central core of an atom when an uncharged particle decays.'

   (ii) 'This type of radiation is particularly dangerous if taken in by the body, as it is heavily ionising.' **[2]**

e) (i) Balance the following nuclear equation.

$$^{\ }_{11}Na \rightarrow {}^{26}Mg + {}^{0}_{-1}e$$

   (ii) State the type of decay which has occurred in this equation. **[3]**

3 **a)** State *two* ways in which ionising radiation may be detected. **[2]**

b) State *two* sources of background radiation. **[2]**

c) Two students carry out experiments to discover the half-life of the radioactive gas thoron. Both experiments were done in *identical* conditions. Their results are shown in the following graphs.

(i) Use the graphs to estimate the half-life of thoron, showing your method on the graphs. Give a result for each student. **[4]**

(ii) One of the students has made an error in carrying out the experiment. State which student, say what the error is and explain how it should be corrected. **[4]**

4 **a)** State one medical application of radioactivity and one non-medical application, other than radiocarbon dating which is mentioned below. **[2]**

b) Radiocarbon dating uses the fact that a proportion of the carbon that makes up living organisms will be the radioactive isotope C-14. Once an organism is dead the proportion of C-14 diminishes at a predictable rate. The half-life of C-14 is roughly 5500 years.

(i) The radioactivity in a living sample of oak is 40 units – what would be the expected activity in the same amount of sample in 11 000 years? **[2]**

(ii) Give a reason why radiocarbon dating becomes increasingly unreliable for organic samples that died more than 30 000 years ago. **[2]**

5 **a)** Ionising radiation produced by radioactive materials can have two undesirable effects on living cells. State and explain what these are. **[4]**

b) Describe how radioactive materials should be stored safely in schools and colleges. **[3]**

6 The diagram below shows a simplified version of the Geiger-Marsden experiment.

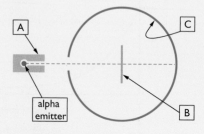

a) Parts A and B are made of metal. State the metal used for each and why it has been chosen for this particular job. **[4]**

b) State the purpose of the coating on the inside of the screen, labelled C. **[2]**

c) Describe what happened to the alpha particles when they reached B. **[4]**

d) What conclusions did Rutherford draw about the structure of the atom from the results of Geiger and Marsden's experiment? **[3]**

7 a) Complete the following sentences about nuclear fission:

Uranium-235 is a _____ material. When an atom of U-235 is struck by a _____ moving or 'thermal' _____ it splits into two lighter elements and emits energy and two or three _____ which, if they strike further U-235 atoms, can start a _____ reaction. An uncontrolled _____ reaction results in a rapid release of enormous amounts of _____. **[7]**

b) The diagram below shows the basic structure of the core of a nuclear reactor.

(i) Label part A of the reactor.

(ii) State the purpose of the graphite blocks. **[3]**

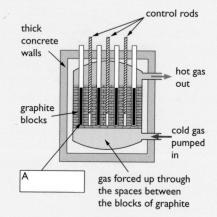

c) (i) If the coolant (the cold gas) flow stops, perhaps because the pumps fail, what will happen to the core?

(ii) An automatic system should detect this change in the core and start shutting down the reactor. Explain what this means and how it is done. **[4]**

Notes

Below you will find some additional material on gravitational forces, states of matter, particle accelerators and smoke detectors that will improve your understanding of these topics.

Gravitational forces

Artificial satellites are also held in orbit around the Earth by gravity. Some have geostationary orbits; these maintain their position above a particular point on the Earth's surface and are used by communication systems and weather monitoring satellites. To be in geostationary orbit the period of the satellite's orbit must be the same as the time for the earth to rotate, one day; the satellite must also orbit the Earth above the equator. Other satellites have different orbits, for example, a polar orbit. Satellites in polar orbits are used for surveillance and mapping the Earth's surface. They have a short period and can scan the whole of the Earth's surface in a few orbits. Polar orbiting satellites have a short working life as their relatively low orbits mean that they are steadily slowed by the Earth's atmosphere, eventually causing them to fall back to Earth burning up in the process.

The states of matter

The changes of state that occur at the melting point and boiling point of a substance involve energy transfers that do not cause a change in the temperature of the substance. For example, if you supply energy at a constant rate to a beaker full of ice and water the temperature will remain at 0°C until all the ice has melted. Once this has happened the temperature of the water will rise until it reaches the boiling point (100°C). The temperature will then remain constant at 100°C until all the water has boiled away. The energy supplied during the process is used in changing the structure of the substance, first breaking the strong intermolecular forces that hold the water molecules together in their solid form, then completely separating the water molecules from one another as the water is converted from its liquid to gas state.

The motor effect

Particle accelerators

A force is produced on wires carrying current in a magnetic field because the moving charged particles in the wire produce a magnetic effect. Beams of charged particles like electrons or protons can be made to accelerate or change direction (or both) by applying strong magnetic fields to them, provided the charged particles are not moving parallel to the magnetic field. This effect is used in high energy particle accelerators like the LHC (large hadron collider) at CERNE and in experimental nuclear fusion reactors.

The applications of radioactivity

Smoke detectors

A type of smoke detector uses an ionisation chamber with a small radioactive source producing alpha particles. Alpha particles ionise (knock electrons out of) oxygen molecules in the air and these enable a tiny current to pass between the charged plates of the ionisation chamber. When smoke enters the ionisation chamber the smoke absorbs and neutralises the oxygen ions causing the current between the plates to fall. This change in current is detected and used to set off the alarm. The amount of radioactive material is very small.

...eration   The rate of increase of velocity with time.

air resistance (or drag)   The force opposing the motion of bodies moving through air.

alpha particle   A type of nuclear radiation consisting of a helium nucleus ejected from an unstable nucleus.

alternating current   A current that continually changes direction.

ammeter   An instrument used to measure the size of current in a circuit.

amplified   Increased in size or power.

amp   The SI unit of electric current.

analogue electrical signals   Electrical signals, usually voltages, that have continuously variable values.

angle of incidence   The angle measured between a ray of light arriving at a surface and the normal.

angle of reflection   The angle measured between a ray of light reflected from a surface and the normal.

balanced   Equal in size but opposite in sign, therefore summing to zero; examples: balanced forces, balanced charge.

becquerel   The rate of disintegration of a radioactive substance; one disintegration per second.

beta particle   A type of nuclear radiation consisting of a high speed electron emitted from an unstable nucleus.

braking distance   The distance a vehicle travels before coming to rest *after* the brakes have been applied.

Brownian motion   The continuous, random, jerky motion of pollen grains observed by the botanist Robert Brown.

cell mutation   A change in the function of a living cell, sometimes caused by ionising radiation.

centre of gravity   The point in a body through which the whole of its weight appears to act.

chain reaction   An escalating nuclear process in which each decay of an unstable nucleus triggers two or more unstable nuclei to decay.

circuit breakers   The modern equivalent of fuses, designed to break the conducting path in a circuit when a set current is exceeded. They may be reset by the push of a switch once the fault causing them to operate is remedied.

comet   A relatively small ice and rock body orbiting the sun with a very elongated (eccentric) orbit. Comets have a distinctive tail.

conductors (electrical)   Materials that allow electricity to pass through them easily. Most metals are good electrical conductors.

contact force   The forces acting on bodies in contact.

control rods   Control rods are used in a nuclear reactor to slow down the rate of nuclear fission or stop the fission process completely.

controlled nuclear fission   An uncontrolled nuclear fission involves the release of vast amounts of energy in a very short time, in short an explosion; in a nuclear reactor the energy is released slowly and usefully by a controlled nuclear fission.

critical angle   Light arriving at a boundary between any material, in which light travels more slowly than in air, and air at an angle greater than the critical angle is totally internally reflected.

current   The rate of flow of electric charge.

density   The mass per unit volume of a substance.

diffraction   The curving of waves as they pass the edges of objects.

digital electrical signals   A digital signal has only two possible values. In computer and communication systems these values are 0 V and 5 V.

displacement   Distance moved in a specific direction; a vector quantity.

distance   Distance moved without considering direction; a scalar quantity.

double insulated   Having an outer casing which is an electrical insulator; having no exposed metal casing.

drag force   The force that opposes the motion of an object through a gas or liquid.

earthed   Having a very low resistance connection to the general mass of the earth, taken as always being a 0 V.

efficiency   A measure of how effectively energy is transformed into a useful form.

elastic   Able to return to its original size and shape after having been deformed.

elastic limit   This is taken as the point that a stretched spring or wire no longer obeys Hooke's Law. This is the limit of proportionality.

electric charge   The property of particles that causes electric effects.

electromagnetic or EM waves   Waves that require no material medium in which to travel. They carry energy as variations in the magnetic and electric fields in space.

electromagnetic spectrum   The family of EM waves, ranging from radio waves to gamma and cosmic rays.

electron   Extremely small particle carrying negative charge and making up the outer 'shell' or 'shells' of an atom.

endoscope   A fibre optic device used to image the inside of living bodies as a diagnostic tool.

energy   Energy exists in many forms – heat, light etc; it is required to do work.

evaporation   The process by which liquids change into gases.

extension   In springs this is the increase in length that results from applying a force to stretch the spring.

fissile   Referring to unstable materials; something that can readily be split or will split spontaneously.

force   A push or a pull. When a force is applied to a body it will cause a change in the state of motion of the body, making it accelerate, decelerate or change direction. Forces can also change the shape of an object.

fossil fuels   Fuels formed from dead organic matter over millions of years; examples are gas, oil and coal. These are non-renewable energy sources.

free electrons   Electrons which are not bound to any particular atom in a solid. These are free to move and enable charge to move through a material forming an electric current.

frequency   The number of waves produced in one second. More generally, how many times something occurs per second.

friction   The force that opposes motion between two surfaces.

fuse   A length of wire designed to melt when a specified current value is exceeded, thus breaking the circuit.

galaxy   A group of many billions of stars rotating around a common centre.

gamma rays   Highly penetrating electromagnetic radiation produced when an unstable atom disintegrates.

geothermal energy   Heat energy produced by nuclear processes in the earth's core.

gradient   The slope of a graph line measured as the rate of increase of the y-axis variable with respect to the x-axis variable.

gravitational field strength   The force in newtons exerted per kilogram of mass by gravity. At the Earth's surface this is approximately 10 N/kg.

half-life   The time taken for half of the atoms in a sample of radioactive material to decay (disintegrate).

hard magnetic materials   Materials that retain their magnetism well.

hydroelectric power   Power produced using the potential energy of water stored in reservoirs in mountainous regions.

hydroelectricity   Electricity produced by generators using hydroelectric power.

inelastic   Materials that are unable to return to their original shape after deformation by a force.

infrared   A part of the EM spectrum. The radiation emitted by hot objects.

insulators (electrical)   Materials that electricity cannot pass through.

joule   The SI unit of energy. 1 joule is the amount of work done (energy transferred) when a force of 1 newton is applied through a distance of 1 metre.

Kelvin temperature scale   The scale of temperature with zero set at the lowest possible temperature that can be achieved: absolute zero. This is $-273°$ on the Celsius scale.

light waves   A part of the EM spectrum that can be detected by the human eye.

longitudinal waves   Waves in which the particles of the medium move backwards and forwards along the same line as the direction of transfer of energy.

loudness   The power or strength of a sound. Loudness depends on the amplitude of the vibrations of the sound wave.

magnetic   Possessing the ability to attract iron and its compounds.

mechanical waves   Waves that require a material medium through which energy may be transferred.

microwaves   A part of the EM spectrum. Used to directly heat water and in telecommunication systems.

moderator   A material used in nuclear reactors to produce 'slow' neutrons needed to trigger nuclear fission. Graphite and heavy water are typical moderators.

moons   Natural satellites held in orbits around planets by the force of gravity.

Motor Rule   The rule devised by Fleming to predict the direction of the force produced on a wire when it carries current in a magnetic field (provided the direction of current is perpendicular to the magnetic field).

negative electric charge   The type of charge possessed by the electron.

neutral   Having no overall electric charge. Neutrons are electrically neutral and atoms are neutral because there is a balance between the number of negative charges on electrons and the number of positive charges on the protons which make up part of the nucleus.

neutron   Uncharged particle found in the nucleus of atoms.

normal   Perpendicular to, as in the normal drawn as a construction line.

normal reaction   A contact force acting at right angles to a surface.

ohm   Unit of resistance; the resistance of a conductor that passes a current of 1 amp when a voltage of 1 volt is applied across it.

optical fibre   A thin glass tube designed to carry information in the form of light through total internal reflection.

parallel circuit   A circuit with two or more conducting paths between any two points in the circuit.

parent nuclide   An unstable nucleus that decays and splits into two or more lighter nuclei. The lighter nuclei are called daughter nuclides.

partially elastic   Description of a collision in which kinetic energy is not conserved after the colliding bodies have separated.

period   The time taken for one complete cycle of an oscillation or wave.

pitch   How high a musical note is. This is related to the frequency of the sound – the higher the frequency the higher the pitch.

planets   Massive objects held in regular orbit around a star by the force gravity.

positive electric charge   The type of charge possessed by the proton.

power   The rate of transfer or conversion of energy.

pressure   Force acting per unit area.

proton   A positively charged particle found in the nucleus of an atom.

radio waves   A part of the EM spectrum. Used in communication and radio and TV transmission.

randomly   Unpredictably.

reaction time   The time taken until there is a conscious response in humans to some event or stimulus.

resistance   A measure of how difficult it is for current to pass through a part of a circuit. Measured in ohms.

resultant force   The net force acting on a body when two or more forces are unbalanced.

Sankey diagrams   Diagrams to represent the relative size of energy conversions that take place in a process or system.

satellites   Man-made objects held in orbit around a planet by the force of gravity.

scalar   A quantity with magnitude (size) but no specific direction. Examples: energy, mass.

second   The base unit of time measurement.

series circuit   A circuit with only one path for an electric current to flow.

soft magnetic materials   Materials that are easy to magnetise and demagnetise.

solar power   Power obtained from the energy transferred by the EM waves from the Sun.

sound waves   Longitudinal waves in gases, liquids and solids with frequencies in the range 20 Hz to 20 kHz.

speed   Distance travelled per unit time.

star   Huge nuclear fission explosions releasing vast amounts of energy as light, heat and other forms of EM radiation.

tension   The force in stretched materials.

thermal radiation   Heat radiation. EM waves with frequencies in the infrared range, lower than the red end of the visible spectrum.

thinking distance   The distance travelled by a moving vehicle in the time that it takes for the driver to react to an emergency *before* applying the brakes.

tidal power and wave energy   Power obtained from the rise and fall of the oceans due to tidal motion and from the waves which result from tidal and wind action on the oceans.

transformers   Electromagnetic devices used to step-up (increase) or step-down (decrease) the size of alternating voltage electricity supplies.

transverse waves   Waves in which the particles of the medium move at right angles to the direction of transfer of energy. Although EM waves do not require a material medium in which to travel, these are also transverse waves.

ultraviolet   The part of the EM spectrum with frequencies greater than the blue end of the visible spectrum.

unbalanced   Not adding up to zero; examples: unbalanced forces have a non-zero resultant or sum.

universe   The system comprising every galaxy.

upthrust   The upward force that acts on an object because it has displaced a volume of liquid or gas.

vector   A quantity that has both size (magnitude) and direction. Examples: velocity, acceleration and force.

velocity   The rate of increase of distance travelled in a specified direction with time.

virtual image   The image formed in mirrors that appears to be behind the mirror. Any image that is not the actual source of real rays of light.

viscous drag   The force that opposes the motion of an object through a liquid.

visible light   EM waves in the range of frequencies that can be detected by the human eye.

voltage   A measure of the energy converted per unit charge passing through a component. Also a measure of the amount of energy transferred to electrical form per unit by an electrical power supply, like a battery.

voltmeter   A measuring instrument for measuring the voltage between two points in a circuit.

volt   The unit of voltage. 1 volt is equal to 1 joule of energy per coulomb of charge passed through a component.

watt   The unit of power equal to a rate of transfer of energy of 1 joule per second.

weight   The force acting on a body due to its presence in a gravitational field.

wind power   Power obtained from the kinetic energy of moving air.

work   The transfer of energy to a body. Mechanical work is the transfer of energy which occurs when a force is applied through a distance in the direction of the force.

X-rays   EM waves in the range of frequencies beyond the ultraviolet range. EM waves that can pass through low density materials like flesh, but which are absorbed by more dense materials like bone.

Index